523.8
C678i
    Cohen, Martin, 1948-
      In darkness born : the story of star
   formation / Martin Cohen. -- Cambridge
   [Cambridgeshire] ; New York : Cambridge
   University Press, 1987.
     p. cm.
     Includes index.
     ISBN 0-521-26270-4

     1. Stars--Formation. I. Title

14 APR 88   12583800 CHCAxc   85-22424

IN DARKNESS BORN

IN DARKNESS BORN

THE STORY OF STAR FORMATION

MARTIN COHEN

University of California, Berkeley and
NASA-Ames Research Center

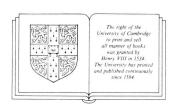

*The right of the
University of Cambridge
to print and sell
all manner of books
was granted by
Henry VIII in 1534.
The University has printed
and published continuously
since 1584.*

CAMBRIDGE UNIVERSITY PRESS

Cambridge

New York   New Rochelle   Melbourne   Sydney

Published by the Press Syndicate of the University of Cambridge
The Pitt Building, Trumpington Street, Cambridge CB2 1RP
32 East 57th Street, New York, NY 10022, USA
10 Stamford Road, Oakleigh, Melbourne 3166, Australia

First published 1988

Printed in Great Britain at The Bath Press, Avon

British Library cataloguing in publication data
Cohen, Martin, *1948-*
  In darkness born : the story of star formation.
  1. Stars – Formation
  I. Title
  523.8 QB806

ISBN 0 521 26270 4

To Barbara, my wife,
For the gifts of love, happiness,
and the inner world

CONTENTS

PREFACE

The roots of this book evolved from a popular lecture about the birth of the stars which has been in great demand since its inception in 1973. Each year, research into star formation has brought fresh surprises. It is my belief that, although astronomy is scarcely a lavishly funded science, we have a duty to keep the public informed about the status of our work. Consequently, this original lecture was constantly undergoing revision. For the past nine years, I have had the opportunity to present an entire course on modern astronomy through "Extension" – the extramural, adult educational wing of the University of California at Berkeley – in which Star Formation takes up three hours of tuition. This course is a non-mathematical, gentle on the physics, descriptive treatment of all the current frontiers in astronomy. I thank the several generations of students who have helped to hone the presentation of ideas in the present book, and Simon Mitton at C.U.P., for providing the external impetus for me to put pen to paper.

There are many colleagues and friends who have observed with me; have spent hours on planes, in airports, on daily commutes, over lunches and dinners, debating and speculating on sundry aspects of star formation. In particular, for their numerous and valuable input, I thank the following: Gibor Basri, John Bieging, David Black, Jesse Bregman, Larry Caroff, Pat Cassen, Mike Dopita, Ed Erickson, Gary Fuller, John Goebel, Mike Haas, Paul Harvey, Dave Hollenbach, Burt Jones, Ray Reynolds, Dick Schwartz, Frank Shu, Steve Stahler, Xander Tielens, Jack Welch and Fred Witteborn.

I thank Wolfgang Kundt for inviting me to be a research visitor at the Astronomisches Institute of the University of Bonn for some weeks where the final chapters of this book were written.

I am grateful to the following journals for permission to adapt figures previously published in their pages: The Astrophysical Journal (Figs. 5.2, 5.7, 5.8, 5.9, 5.10, 5.14, 6.7, 6.8, 6.11); The Astronomical Journal (Figs. 6.2, 6.3, 6.4); Monthly Notices of the Royal Astronomical Society (Figs. 5.3, 5.15, 10.2).

Finally, I am deeply indebted to my wife, Barbara, for helping me to coax the original manuscript into a word processor during our all-too-heavily committed weekends; for her understanding while I flew hither and yon to observe young stars; and for her active collaboration during long evenings when I was absorbed by writing, revising, draughting diagrams, and proofing this book.

Oakland, California   Martin Cohen
September 1987

1

What does astronomy tell us?

Say something to us we can learn.
By heart and when alone repeat.
Say something! And it says "I burn."
But say with what degree of heat.
Talk Fahrenheit, talk Centigrade.
Use language we can comprehend.
Tell us what elements you blend.
Robert Frost
"Choose Something Like a Star"

Their height in heaven comforts not,
Their glory nought to me;
'Twas best imperfect, as it was;
I'm finite, I can't see.
Emily Dickinson

Problems of astronomy – stellar properties and evolution – luminosity and brightness – colour and temperature – temperature and radiation – the electromagnetic spectrum – wavelength and frequency – fingerprints – astronomical yardsticks

---

1.1. Introduction

What is astronomy? Astronomy provides the means to answer all those simple, big, fundamental questions about which everyone is curious. What is the Universe? How big is the Universe? How old is it? Where did the sun come from? How did the planets form? How did we get here? What are stars? Where are the stars? Astronomy is the framework of specialised physics within which all these questions have meaning and can be addressed. You can't be curious at all without wondering about questions like these. They're all so obvious; so easy to ask. The issues may be specialised but they're important to all of us. The very materials of which you're made were assembled and cooked up inside a particular kind of star. The mere fact that you're sitting on a planet, reading this, whilst that planet circles a star is sufficient testimony to the validity and relevance of astronomy.

This book is designed to answer some of these simple questions – not all, for that would take too long and raise too many other questions along the way. But there will be questions aplenty, and issues to be thought about. I hope that you will have a much clearer impression of the sun's origins and perhaps of your own after reading it. My aims will be to show you what it takes to make a star like our sun, and to sketch out ways that the manufacture of a star can lead to strange behaviour and to debris out of which planets can eventually form. Some of these planets may even synthesise, harbour and sustain life forms but that is not the province of the astronomer – that's where the equally specialised tools of the extraterrestrial biologist, or exobiologist, take over from our own investigation. Our prospectus then is to take a global view of our Galaxy; to find those regions that can be identified as the probable birthplaces of stars; to look at young stars by a host of techniques so that we can recognise how they differ from ordinary mature stars; to go back in time as far as we can in the story of an individual star; and to approach the beginnings of stellar existence. There will, I hope, be many surprises, for stars are remarkable entities, deserving of our most careful examination. When our journey eventually ends it will be important to remember that we will not have exhausted all the phenomena of which stars are capable, nor will we have "seen" the precise moment of birth of any star. Rather we will have constructed a broad conceptual framework into which we can fit all the phenomena of which our latest observations and theories have made us aware.

Science is a continuously growing structure. There is no "right" theory – there are only wrong ones, that fail to explain, predict or at least accommodate known facts. Fields of science very rarely stop, or reach a point where theory and observations are in complete accord. There is always something new to learn; a new style of behaviour to investigate; new facts to be gleaned by different techniques; promising starts to be abandoned in favour of what once seemed far-fetched notions.

Science is dynamic. The stars will take us as far as our minds will let us go – to the next horizon, the next problem, the next challenge.

1.2. Problems of astronomy

Look up at the sky on a dark clear night and try to imagine that all the stars you can see must be placed in their correct relative positions; some close to us, some at extreme distances. This range-finding procedure is not an easy problem to apply to individual stars. Imagine the difficulties magnified when we take a deep photograph of a dense starfield (Fig. 1.1). Therefore, how are we to know with exactly what kind of a star we are dealing? The sky is not a tangible environment like a laboratory; we are not permitted to make several trial experiments on real stars to assist us in our understanding of what they are and how they function. Add to these constraints the seemingly constant nature of most of the stars that we observe, yet recognise the necessity of putting the different types of star into some kind of evolutionary scheme; for they do evolve.

The life of an individual astronomer is less than a century. The collective consciousness of the human race does not much exceed one hundred thousand years. Our own sun is five thousand million years old. We need to know what it looked like when it was in its infancy, and to understand how it became the star we see today. It is sometimes hard to believe but the stars are not fixed entities, constant and reliable as Shakespeare would have them. They bear a superficial resemblance to creatures such as ourselves. First, there exist parents – "dark clouds" of dust and gas – from which stars are born. A period of early turmoil establishes these "protostars" on their evolution to maturity. Long æons of stability characterise the middle years of stars until internal catastrophes (of a type pre-determined essentially at the moment of birth) return the materials that once were bound into the former stars to the ever-present gaseous ocean that floods the Galaxy – the "interstellar medium". There is even a recycling process to stellar existence whereby each generation of stars contains, as seeds for its formation, matter that was ejected from previous generations of stars. The principal

problem for the astronomer is not the lack of continuous evolution
of stars but rather the very long periods of time required for even

Figure 1.1   A busy starfield showing many of the ingredients with which
we shall deal: individual bright young stars; glowing clouds of gas; dark
obscuring patches of dust; and tiny dark globules silhouetted against the
bright nebulæ. (Curtis Schmidt photograph: reproduced with permission
of the National Optical Astronomy Observatories. Use of this photograph
does not imply the endorsement by NOAO of any individual, philosophy,
commercial product, process, or service.)

the fastest stellar developments. Ten thousand years is a cosmic instant but, for us, perhaps the longevity of a civilisation.

Imagine you represent an alien intelligence, brought to earth for a total period of one day and taken briefly to view several institutions relevant to human activities and development. Your day begins in a maternity ward where you witness the emergence of new small life forms from larger ones. In a kindergarten you see a clustering of small creatures around a larger life form, an apparent symbiosis which persists through secondary school and into university life. After a short observation of the process of government you end your day with a visit to a home for the elderly, and a fly-by of a cemetery before leaving this planet. You have not observed the entire evolution of a single human being, for one day is a tiny fraction of the lifespan of the average person. Nevertheless your assignment is to organise these fleeting "snapshots" of different humans, in a variety of phases of life, into a coherent picture of individual human life from conception to death. Your task has, of course, been greatly facilitated by the thoughtfulness of your guide who caused you to see these snapshots in their correct sequence. But, as aliens, you may not even realise that this sequence was a chronological one.

The astronomer has no such thoughtful guide beyond physics and his or her own intuition, but the analogy is otherwise precise and useful. We see many snapshots of stellar life around us but which represent the earliest and which the latest phases? Are we certain that, if we do attempt a chronology, we have not stuck the supposed birth of one type of star on to the maturity of a totally different type? It is the role of astronomy ultimately to provide us with a framework suitable for interpreting all kinds of stellar behaviour. The construction of this framework is based upon two types of input. Observations are crucial; phenomena must be categorised, given names for discussion; stars must be placed in appropriate bins. By these means it is possible that we may discern patterns in behaviour. From the recognition of these patterns comes physical understanding of the phenomena. In a parallel development, theories are draughted on the basis of our knowledge of physics and these theories must incorporate old observations and respond to new ones. Most important, theory must try to predict other forms of behaviour, either unobserved as yet or at least unrecognised. Similarly, observations must respond to the consequences of specific mechanisms: are there definitive tests of hypotheses that can be devised?

Under ideal circumstances, the development of theory and observation should be simultaneous, with copious intercomparison and dialogue. Sometimes the advent of new technology makes possible the design of a new type of computer suitable for following, in unprecedentedly fine detail, the life of a theoretical star. A comparable breakthrough in technology might enable a specific style of observations to be undertaken on a large body of stars where previously only a few objects could be studied by this same technique. Or, equally crucially, new sensors may open up to observation a part of the spectrum not previously accessible. At the time of writing, our observations have undergone an explosion. Even within the once very restricted field of "star formation" we have been presented with a plethora of diverse facts and patterns, most within only the past very few years. The challenge now is to organise these facts into meaningful patterns; structures that will define the next generation of theories. It is at this exciting stage that you meet the topic of the birth of the stars, perhaps for the first time. It may seem to be only a minuscule piece of astronomy. If so, then use this feeling to appreciate the vastness of the context of astronomy in its entirety. Do not misinterpret the smallness of our field to imply its insignificance, for, to understand our own existence and that of our planet, of our solar system, depends upon a firm grasp of those processes that attended the birth of our sun.

1.3.   An easy way to look at stars and their properties

Let us return to the night sky. What are the principal characteristics of the different stars that you detect with your eyes? Only two are necessary to describe your perception of these tiny points of light: their colours and their observed brightnesses. This may seem to be a very simplistic description of systems as complicated as one might imagine stars to be. However, it suffices to reveal some instructive patterns which offer an opportunity to do more than merely sort the stars into types. For example, if this exercise were applied to human beings, one might select height and weight as a crucial pair of observational attributes. One would also have some expectation that a trend should emerge from a comparison of these two, a priori quite independent, quantities. By and large, the tallest people would be the heaviest, and the shortest would be the lightest (Fig. 1.2). Now we know that all these people represent the same phenomenon – all are life forms of a common species – and that none of them sprang into existence, like Pallas Athena, in an instant. We realise that the relationship between these two parameters can also reflect evolution: people enter the world as small, lightweight entities and, during later years, increase both in size and mass.

Further, in the first months and years of a baby's life, the increase in weight and height is prodigiously accelerated compared with that in any other phase of normal life. Similarly, the diagram of height versus weight encapsulates a statement as to how frequently one encounters people of different weight and height: there are relatively few extreme cases at the upper end of the trend line, and rather more at the lower end, especially in times of "baby

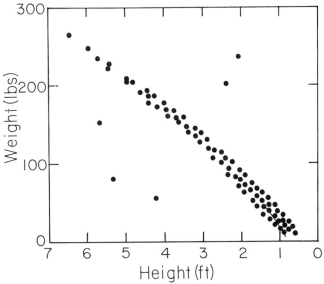

Figure **1.2**   Plot of two basic human attributes – weight and height. Note the path through the diagram along which most people lie, and the little evolutionary track followed by an embryo.

booms''! All these observations and our knowledge of the evolution of humans can be captured in a single diagram representing the dependence of weight upon height. We can even explain the locations of a number of points high above the average trend: these signify obese people, overweight for their height. We recognise that this phenomenon is not uncommon and that these points, too, are drawn from the same population to which we all belong. The birth and subsequent evolution of a child can be represented by a continuous track, pursued very rapidly at first, and thereafter much more slowly.

Likewise, early astronomers sought to investigate the relationship between the two empirical stellar variables, namely colour and brightness. Such a diagram (Fig. 1.3) is strikingly similar in its superficial details to the one we have just produced for humans. What could the diagram tell us? How are stars in one part of the plot related to their neighbours? There is a broad swathe across the diagram, representing the most typical members of the stellar population. This rather fuzzy locus is termed the "Main

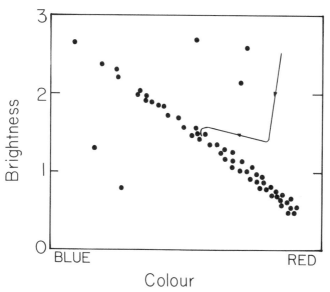

Figure **1.3**   Stellar analogue of Fig. 1.2: brightness versus colour plot. The main sequence is the locus on which typical stars lie though many more dim red stars are found than bright blue ones. The track represents the pre-main-sequence evolution of a star like our sun.

Sequence", for it was originally felt that most stars, in some manner, travelled in time along this path. The density of stars along the main sequence far exceeds the sprinkling of stars elsewhere in the diagram; we also see that the upper end of the sequence contains far fewer points (stars) than the lower. As for humans, we find a scattering of stars above the sequence – stars that appear to be overly bright for their colour.

It is worthy of note that this plot was first constructed by Ejnar Hertzsprung in 1911 and, independently, by Henry Norris Russell in 1913, hence its popular name – the "Hertzsprung-Russell diagram" (we shall find it convenient to shorten this to HR diagram hereafter). Although the broad locus was originally thought of as an evolutionary sequence, astronomers have since recognised that this is not a correct interpretation of the diagram. Rather it represents the importance of the mass with which a star is born; but we shall defer discussion of this aspect of the HR diagram until Chapter 2. In Chapters 2, 5 and 6 we will investigate the evolutionary tracks of young stars in some detail. For the present we represent these by a single track, which our sun is believed to have followed in its formative and adolescent years. This track, albeit of different shape, is the analogue of that of the baby in Fig. 1.2. This analogy does not connote anthropomorphism! Rather, the concept of a variable rate of evolution is common to both humans and stars, and this may serve as a helpful analogy.

1.4. Luminosity and brightness

We must now examine the variables plotted in Fig. 1.3 more carefully, to clarify the concept of "luminosity" and to develop the relationship between the somewhat imprecise variable "colour" and the quantitative and physically useful "temperature". There is a similarity between the two quantities "weight" for humans and "brightness" for stars. Our "weight" is a response to the gravitational pull of the earth: transport us to Jupiter and we would be flattened by our "weight" on that planet. Weight is not an absolute quantity – it depends upon the locale in which it is felt or measured. But it also depends on an absolute quantity; we call the absolute attribute of the colloquial term "weight", the "mass" of an object. Suppose, for example, that we wished to replot Fig. 1.2 a millennium from now, when the human race may have spread to far-flung planets across the Galaxy. Different planets would have different values of gravity rendering even the same person's weight a function of his environment. However, plotting mass against height would produce a diagram with intrinsic value throughout the Galaxy. (Note that height is already an intrinsic quantity, give or take relativistic situations in which it can alter.)

What of stars? We previously used "brightness" as the ordinate (the vertical axis) in Fig. 1.3, but this too depends upon our perception of the stars. Two astronomers may have eyes with differing responses and acuities to colour. This ambiguity can be alleviated by giving everyone the same coloured glass filters through which to look at the sky. A more serious problem arises when we think about the sun. Where does the sun lie in Fig. 1.3? Clearly we perceive the sun to be the dominant source of all radiation on earth. The next brightest stars are seen as mere pinpoints in the night sky, and are mostly too feeble to be seen during the daytime. If we were to travel to the nearest star beyond the sun, and to look back at the sun, it too would be very faint and our sky would now be dominated by a different nearby star. To extract physically meaningful statements from the HR diagram, the role or location of the observer must be unimportant to the determination of variables. That the sun is so "bright" is merely fortuitous: it is the closest star to the earth. What is needed is an absolute quantity so that all stars can be compared fairly, without concern for their proximity, or otherwise, to us. Astronomers, therefore, use "luminosity", which is a measure of the total energy radiated (as seen through our standard glass filters) by the surface of the star. To relate luminosity to brightness requires knowledge of the distance of a star from the earth. If two stars have the same brightness, yet one is ten times more distant from the earth, then the more distant star has 100 times the luminosity of the fainter. This is an example of an "inverse square law"; that is to say, the brightness of a star, of fixed luminosity, is perceived to diminish as the product of the distance with itself (the "square" of the distance) increases. To make this more concrete, let us compare the stars

Vega (the brightest star in the northern sky, in the constellation of Lyra, the Lyre) with Deneb (the brightest star in Cygnus, the Swan, or the Northern Cross). Both have closely the same colour (blue-white) and, very roughly, the same apparent brightness. Yet Vega is only 26 units of distance ("light years": see Chapter 1.9 below) away whereas Deneb shines from almost 1400 units! Consequently Deneb's luminosity exceeds that of Vega by a prodigious amount – a factor of almost 3000 – yet their brightnesses are comparable from our viewpoint.

1.5. Colour and temperature

Now that we have our intrinsic ordinate for the HR diagram it is appropriate to consider the abscissa (the x-axis, or horizontal coordinate). Previously we introduced the concept of standard glass filters, intended to reduce, or even eliminate, subjective differences between observers. However, we are seeking another absolute, or intrinsic, quantity, the stellar temperature. How does this relate to colour? Many readers, even if only slightly familiar with the night sky, may have a better understanding of this relationship than they realise. What colours do we perceive some of the brightest naked-eye stars to have? Let's begin with, hopefully, the best-observed star – the sun. It has a surface temperature of about 6000 degrees (in units to be discussed later: Chapter 1.6) and it has a yellow appearance. What of Sirius or Vega, both strikingly bright stars – the former, the brightest in the sky (apparent brightness, remember), the latter, the brightest in the northern celestial hemisphere? These are stars much hotter than the sun, almost 10,000 degrees at their surfaces. To our eyes they seem white or bluish-white. Finally, think of winter in the northern hemisphere, when the wonderful constellation of Orion strides across the sky. At Orion's right shoulder lies the star Betelgeuse – a very bright, distinctly orange star. Betelgeuse is actually very much more distant than the other stars we've been thinking about, but its surface temperature is all we need for now: that is about 3000 degrees. These stars have been ordered in Table 1.1 to constitute a sequence of representative colours and corresponding surface temperatures. Now we can see how the colour becomes redder as the surface temperature is lowered. This is not to say that all the sun's light is yellow, nor that Sirius produces only blue radiation. But the dominant impression that we receive from each of these stars makes a statement about what colour represents most of the radiation; that is, where in the spectrum lies the peak of the visible radiation from each star.

1.6. Temperature and radiation

It is valuable to consider in a little more detail what stellar radiation looks like, and how this relates to our visible wavelengths. Let us observe the three stars of Table 1.1 through two standard glass filters called "blue" and "red". As ordinate we use the intrinsic energy radiated by a star per unit of surface area; horizontally we plot the colour of the radiation. Our filters are represented in Fig. 1.4 by the cross-hatched vertical bands which indicate the slice of spectrum that each glass transmits. Thermal physics tells us the shape of the curve that represents the radiation emitted by ideal objects, characterised by temperature and by the statement that these ideal radiators are as efficient as it is possible to be in emitting at each wavelength. We call these ideal radiators "blackbodies" and we style their spectra "blackbody curves", each labelled solely by a single parameter, the temperature. If our three stars had identical sizes and were equally distant from us, and we further assumed them to be blackbodies, then the three curves in Fig. 1.4 would represent their radiant energy curves.

Two attributes of these curves are immediately obvious. First, as our eyes tell us, hotter stars emit mostly blue radiation, cooler stars mostly red radiation (see the horizontal positions of the peaks of the curves). Secondly, cooler stars always emit less energy, per unit area of stellar surface, than do warmer ones, at any wavelength. (So

Table 1.1 *Stellar colours and surface temperatures*

Star	Colour	Surface temperature
Sirius	bluish-white	10,000 degrees
Sun	yellow	6,000 degrees
Betelgeuse	orange	3,000 degrees

the curves for hot stars always lie above those for cooler stars.) To define the peak of such a curve would necessitate observations with many different coloured filters but astronomers have very finite lifetimes. What can be achieved with only our two, red and blue, bands? See the relationships between the amounts of energy that each star emits in these bands. Our hot star produces vastly more blue than red radiation; our cool star shows exactly the opposite behaviour. A star with intermediate temperature would appear almost as bright through both filters. Suppose we were to define "colour" quantitatively to mean the ratio of energy received through the "blue" band to that received through the "red" band. Then this "colour" would decline steadily with decreasing stellar temperature. Finally we have achieved our quantitative measure of temperature, and have related it very simply to the idea of colour.

You may be wondering how to calculate the actual amount of energy radiated by a star since there are obviously going to be big stars and little stars; you would not expect these to radiate the same

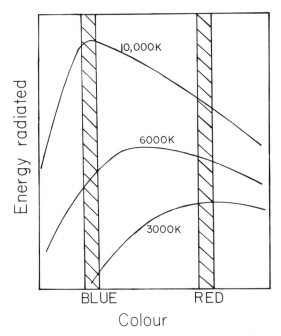

Figure **1.4**   Energy radiated by three stars of the same size but different temperatures. Cross-hatching denotes our blue and red filters.

quantity of energy even if both had the same temperature. Just think of what you see when you look at the sun: you see a disk, with a well-defined area. So, for any other star too, what counts is the the area that we see projected in the plane of the sky; in other words, the area of the apparently flat circular disk, not the actual area of the curved surface of the star. What else might the amount of radiation depend upon? We have already mentioned the obvious parameter, the temperature. But how does the temperature enter into this calculation? It is a crucial parameter that greatly differentiates cool and hot stars from each other. Temperature enters as the fourth power. In other words, if we compared two stars of the same radius but with temperatures of 3000 and 6000 degrees, then the hotter star would radiate 2 times 2 times 2 times 2 (the "fourth power of two") times the radiant output of the cooler star, or a factor of 16 times as much.

Incidentally, what are these "degrees" that we have introduced to describe stellar temperatures? Are they the everyday Fahrenheit or Celsius degrees with which we are all familiar? No. Such scales are not the most convenient in physics for which we use "absolute temperatures", in accordance with our desire to use preferentially intrinsic quantities (ones that don't depend upon human conventions). How do we make ordinary temperatures into absolute ones? Physics describes a lowest temperature, below which nothing in the Universe can be cooled. We shall discuss atoms and electrons later, in Chapter 2.5. For the moment we note only that everything in the Universe is in motion; electrons are always whizzing around in atoms; atoms are constantly jostling about in larger structures like desks and ice cream cones. Why? Because everything is at some non-zero absolute temperature. The higher the temperature, the greater the scale of these random atomic motions. What would happen if we were able to achieve absolute zero? Every motion would halt. All the hurly-burly of particles would drag to a stop. Nothing would be moving by virtue of random motions, of "thermal motions", as we term them, noting implicitly a relationship between temperature and atomic motions. This should anchor the conceptual existence of absolute zero. Its value relative to the everyday scale is equivalent to -273.12 C, a chilly level denoted by 0 K ("K" stands for Kelvin, after a nineteenth century physicist famous for his work in thermal physics). A convenient yardstick is that our usual environment is at roughly 300 K, and one degree Kelvin represents the same step in temperature as one degree Celsius. For stars, it makes little difference whether we use Celsius or Kelvin scales: 5700 C and 6000 K both seem inconceivably hot! But the strictly correct units, that render valid our discussion on blackbodies, and which we shall use consistently in this book, are degrees Kelvin (or degrees absolute, as they are sometimes called).

1.7. The electromagnetic spectrum: the broad view

Are we now familiar with all the nuances of the HR diagram? Not yet, for there is one serious problem to be resolved. We have seen that two coloured filters will suffice to estimate stellar surface temperature by way of "colour", namely the ratio of energies perceived through each filter. Luminosity we recognise as

Table 1.2 *The electromagnetic spectrum*

Radiation	Typical Frequency (cycles/sec)	Typical Wavelength	Platform for observations
Gamma rays	1(22)	0.0003 Å	Satellite
X-rays	1(19)	0.3 Å	Satellite Rocket
Ultraviolet	3(15)	0.1 micron	Satellite Rocket
Visible	6(14)	0.0005 mm	Ground
Infrared	3(13)	0.01 mm	Ground Aircraft
Far-infrared	3(12)	0.1 mm	Aircraft Balloon Satellite
Radio	3(9)	10 cm	Ground

Notes:
3(12), for example, represents 3 followed by 12 zeroes.
"Å" denotes Angstrom unit, a measure of length equivalent to one hundred-millionth of a centimetre.
A micron is one millionth of a metre, or one thousandth of a millimetre.

an absolute version of brightness, taking into account stellar distances. But how are we to know how much luminosity is emitted in the form of radiation to which our eyes are not sensitive? Looking at Table 1.1 there seems to be no reason why 1000 K and 100,000 K stars should not exist but, if they did, what would they look like to us? The very cool (1000 K) star would emit no perceptible blue radiation and would peak so far into the red that it would be almost a pure "infrared" source, imperceptible to our retinas. Similarly, the very hot star (100,000 K) would have its radiant peak far into the ultraviolet and even our bluest view of it would considerably underestimate its true luminosity. It is not sufficient to study the Universe solely at our visible wavelengths: there is a richness to the cosmos that can only be recognised by becoming less parochial in our observations. There is nothing magical about the tiny sliver of the electromagnetic spectrum that our eyes happen to see best. Entire and substantial components of our Galaxy would have remained unsuspected without radio, infrared, ultraviolet and X-ray observations, for there are sources of electromagnetic radiation that emit either entirely, or at least dominantly, in regions of the spectrum "invisible" to us. Table 1.2 expresses the entire panoply of electromagnetic radiation of which we are aware, showing wavelengths and indicating the requisite methods of observation.

Light, indeed all electromagnetic radiation, may be represented as a series of waves with regular up and down variations (Fig. 1.5). The "frequency" of light dictates how many complete cycles (from

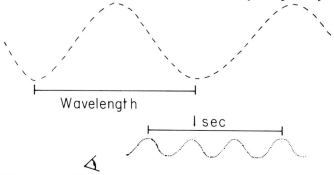

Figure 1.5   Wavelength of an electromagnetic wave and frequency, the latter represented by the number of complete wave cycles that pass the observer in one second.

wave crest, to trough and back to crest again) of the wave pass a fixed point in one second. Its "wavelength" is a direct spatial measure of the length of this wave – from crest to crest, or trough to trough. An implicit relationship exists between frequency and wavelength for any wave motion. The product of the two is a constant, equal to the velocity of the wave. So wavelength and frequency are inversely proportional: radio wavelengths are long, even graspable (inches and feet in scale), hence their frequencies are very low; light is radiation at an exceedingly high frequency, and its wavelengths are correspondingly small (twenty millionths of an inch represents the wavelength of green light). Incidentally, all electromagnetic radiations travel at the same speed – "the speed of light" – 186,324 miles per second (300,000 km/sec), or 7 times around the earth every second!

It is hard to be a cosmopolitan observer because some forms of radiation cannot penetrate to the surface of the earth and many require balloons, high-flying aircraft, even rockets or satellites for their detection. Our lives are circumscribed by many different forms of radiation. Even heat is electromagnetic in origin and is carried by infrared radiation: imagine dropping the temperatures in Table 1.1 below even 1000 K, towards the kinds of temperature range that are common in daily life. These "cool", by stellar standards, but hot by our experience, blackbodies would clearly be emitters of infrared energy and of little else. Only sound – that we think of as "travelling in waves" or being "radiated" – is not electromagnetic in nature, although it is still a wave phenomenon. The acid test is that only electromagnetic radiation will propagate through a vacuum (sound depends on the jostling of molecules and atoms that lie between sources and our ears for its perception).

Since we have also stressed the value of comparison between theory and observation we should be aware that a theoretician, calculating the characteristics of a star, would naturally consider all sources of radiation, and would not wish to be restricted by our limited notions of visibility. We need to refine our concept of luminosity to include all forms of energy: this we term "bolometric luminosity" (from the Greek word "bole" meaning a ray; hence the bolometer is a very sensitive device for measuring any radiant energy without limiting ourselves to any specific part of the electromagnetic spectrum); this is the final form of the ordinate that we shall plot in our HR diagrams. This is not to say that every star does emit copiously in all parts of the spectrum but that we should be prepared to measure more than just visible radiation.

1.8.   Fingerprints and the interstellar detective

You may be wondering what would happen if something were not an ideal blackbody radiator. In reality, few things in the Universe do radiate like blackbodies: most objects cannot radiate with 100% efficiency at all wavelengths. Indeed, most things don't even radiate with the same efficiency at all wavelengths. This, of course, makes the Universe a more complicated place to study but it does have the merit of "leaving fingerprints" for astronomers to recognise. What I mean by "fingerprints" is that different substances have peak radiating efficiencies at different wavelengths. This fact is of great value in the infrared regime and it enables us to distinguish at least broad classes of material. Fig. 1.6 illustrates the efficiency of radiation of two different materials at infrared wavelengths (the units of wavelength are "microns": 1 micron is one millionth of a metre), as measured in a laboratory. Note the way that each has a different "signature" – the shapes,

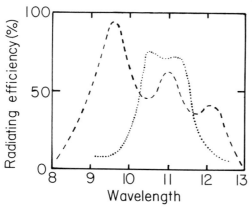

Figure **1.6**   Schematic curves that represent the radiative efficiency (or "emissivity" as it is sometimes called) as a function of wavelength for two types of dust particle. Dashed – silicates; dotted – carborundum. The wavelength is in units of "microns" or "micrometres" and signifies infrared wavelengths some twenty times as long as yellow light.

wavelengths and relative peaks of radiation efficiency curves differ. These curves suffice for us to identify, albeit rather broadly, of what approximate chemical composition were the materials that produced this infrared radiation. The astronomer can play interstellar detective, hunting down the specific materials responsible for these fingerprints. Sometimes he finds these signatures "in emission" in space – like pale, greasy fingerprints on a dark pane; sometimes "in absorption" – like prints on an otherwise dusty table. The method is still the same and the culprit is always recognisable.

We shall meet the idea of physical fingerprints again in Chapters 3 and 10; they are an important weapon in the astronomer's arsenal, and are not confined to the infrared spectral region. In fact, quantum mechanics (Chapter 3.2) entitles us to play detective even at optical wavelengths and to read and comprehend the most basic messages conveyed by starlight.

1.9. Astronomical yardsticks

Your intuition may tell you that space is very big, perhaps even that it is vast beyond comprehension. If so, then you will have realised that puny units like miles are incomparably too small to tackle the description of celestial distances. One handy unit, adequate for parochial purposes, is the average distance of the earth from the sun ("average" because our orbit is not circular: see Chapter 2.3). This is 93 million miles. It is called "the astronomical unit", or abbreviated to "AU". We can use it to describe Pluto's mean distance from our sun as about 40 AU, or to state that if there were another, as yet undiscovered, planet in our solar system, even more massive than Jupiter, it could lie 100, 1000 or even 10,000 AU from the sun. But the nearest known extrasolar star is already over one quarter of a million AU from us. Consequently, even the AU is worthless for practical stellar distances.

Another "natural" unit derivable from the earth's orbit is the "light year": namely the distance a light ray would cover during one revolution of the earth around the sun (1 year). This is 5.7 million million miles and as such is approaching usefulness. That nearest star beyond the sun – Proxima Centauri as it's often styled

– is then 4.25 light years (LY) from us. As we have already mentioned, some of the apparently bright stars are also close to us, like Vega at 26 LY. Of course, there are lots of stars that lie thousands of LY from the sun. For these it is more practical to jettison the LY and replace it by another unit, that also depends upon the orbital characteristics of the planet upon which we happen to find ourselves. This unit is the "parsec" (pc). It's a contraction of the words "parallax" and "second" and it refers to the distance from the sun at which the radius of the earth's orbit (1 AU) subtends an angle of 1 second of arc (1/3600th of 1 degree). A parsec is roughly 3.26 LY and the convenient thing about parsecs is that astronomers have chosen to use prefixes to qualify them. So we have "kpc" for "kiloparsec", equal to 1000 pc; and "Mpc" for "Megaparsec", or one million pc. Armed with the LY and "the parsec family" we are ready to measure the cosmos!

1.10. Synopsis

With these very basic concepts at our fingertips we are ready to consider the inner mechanisms of stars and the dependence of their evolution upon mass (Chapter 2). The HR diagram will continue to be a valuable interpretative tool. In Chapters 3 and 4, the focus switches from stars to their birthplaces and progenitors – the dark clouds that are parents to new generations of stars and which fulfil the role of stellar nurseries. Many strange phenomena attend the birth of stars, even of relatively small stars like the sun, and we shall be concerned with the nature and significance of these events in Chapter 6. They affect both individual stars and the surrounding dark clouds. The theoretical journey from cloud to star is addressed in Chapter 7, after we have compared observations of actual young stars (Chapter 5) with theoretical predictions. Armed with a picture of the vigorous early years, we shall plot the subsequent evolution of stars towards the main sequence in Chapter 8, the end of our journey. There are several controversial issues that we shall explore in Chapters 9 and 10. These include the possible necessity for some process to initiate the collapse of clouds to stars; potential differences in the ways that massive and lightweight stars may form; the conditions of the early solar nebula and the formation of planets, at least insofar as it relates to what we can learn by looking at young stars; and the origin of the first stars which will elevate our view again from the parochial to the Galactic.

2

What is a star?

Art thou pale for weariness
Of climbing heaven, and gazing on the earth,
Wandering companionless
Among the stars that have a different birth, - ?
P. B. Shelley, "To the Moon"

One star differeth from another star in glory
The First Epistle of Paul to the Corinthians, xii. 41

*Stars and planets - energy - stellar collapse - nuclear fusion -
the smallest and largest stars - mass and evolution of stars*

2.1. Stars and planets

One of the most popular astronomical misconceptions relates to the differences between a star and a planet. A casual observer of the sky, unfamiliar with the normal groupings of stars - the patterns we term the "constellations" - will find it difficult to distinguish a star from a planet. They will notice, when Jupiter or Mars, for example, is pointed out to them, that the planets are usually among the brightest starlike objects in the sky. Over a period of time they could detect the motion, with respect to the constellations, of the brighter planets such as Venus or Jupiter. This explains the name "planet", derived from the Greek word for "wanderer"; for ancient astronomers readily observed that a few celestial lights were not confined to specific constellations but moved through these. An easy test, that will usually suffice for the distinction between stars and planets is that stars twinkle while planets do not. Why should this be the case?

To our naked eyes, planets appear every bit as small, as pointlike, as "unresolved", as do stars, yet even the unaided eye can recognise that this is not true. Stars twinkle because they are so far away as to be effectively mere points: the rays of light that they emit appear to derive from a single point. This singlemindedness is a disadvantage when passing through our atmosphere. The constituent atoms, molecules, specks of dust that make up our planet's thin, but to us crucial, gaseous skin readily scatter, bend, absorb and break this single ray up into many less well-directed beams. Consequently not all the energy radiated by a star reaches us. Entire chunks of the stellar spectrum are lost to the atmosphere. These suffer absorption, leading to the heating of the atmospheric gases and molecules. Other radiations, notably light, are attenuated minimally by absorption but are routed into many different paths, not all of which will meet again at our eye. We perceive the ensemble of these scattered rays that from moment to moment finds its way to our eye, but the brightness of this group of beams is not due to the same collection of differently scattered rays. At one instant, many rays will strike our eyes and blend together; at the next, almost none do so, and the star will seem to "twinkle" - to fluctuate erratically in brightness. If you are perceptive you will notice that this twinkling is not a careful preserver of stellar colour. You can see Sirius, a basically blue-white star (its true colour), flash from green to yellow to blue, occasionally even to red, quite erratically. Even Betelgeuse, essentially a red-orange star (intrinsically) can occasionally sparkle blue. This happens because the atmosphere not only causes light rays to lose their memory of direction but also "disperses" light: that is, it refracts rays (bends them), as does a glass prism, to an extent that depends on the colour (wavelength) of light. So white light is separated into its constituent colours, all of which are then individually subject to atmospheric scattering.

The existence of wavelength-dependent phenomena such as scattering and dispersion is actually a matter of daily experience. A clear sky is blue because scattering of blue light is far more serious than that of red. Look at the sun high up in the sky: much more blue radiant energy is diverted from the direct route to your eyes than is red energy. Consequently, blue rays are more frequently to be found meandering aimlessly through the atmosphere; eventually, some are detected by our eyes in any random direction. So "the sky" is blue. By contrast, as the sun drops in the sky, our line of sight penetrates through a very great depth of atmosphere compared with the short noontime path (Fig. 2.1). The light now encounters a greatly enhanced number of scatterers (principally molecules of nitrogen and oxygen), further diminishing the amount of relatively directly transmitted short wavelength (blue, green and even yellow) light. So the setting sun appears much redder (lacking in blue) than its customary colour. An even lower sun encounters another process - bending or "refraction" of light along this very long gaseous pseudo-prism. In fact, even when the sun's disk is slightly below the horizon geometrically, we can still see its light rays (Fig. 2.2). At this time, some light can reach us by refraction. By these means the interplay of refraction, dispersion, and scattering accounts for the wonderful richness of colour that attends the setting sun. If you want to know just how tough the life of blue light really is, compared with that of red, the scattering of blue light is 12 times greater than that of the deepest red light that our eyes can detect.

While we're thinking about the sun, which is itself a star, does it twinkle? Life on earth would be exceedingly irritating if it did to any perceptible extent, but why should it not sparkle like the other stars? The reason is that we can "resolve" the solar disk; that is, we can distinguish a host of separate points on its surface, each of which is twinkling all the time. But the addition of bundles of light rays from so many individual points, each bundle suffering its own random scatterings, produces, on average, a steady output. Our

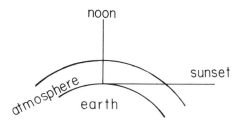

Figure 2.1   At noon the path of sunlight through our atmosphere is shortest; at sunset it is longest.

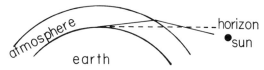

Figure 2.2   The sun's rays can be refracted to our view (solid line) even when the sun is geometrically below our horizon (dashed line).

sun shines at a constant level as far as our senses can determine. We should therefore amend our stellar discriminant: all stars, except for the sun, twinkle as seen from the surface of the earth. As we look at more and more distant objects, their apparent sizes shrink until they too are points. "Twinkling" then is not an intrinsic property of luminous objects but rather a statement about the presence and properties of our atmosphere. Any celestial object, close enough to us to have a perceptible diameter, will not twinkle. That is the reason for the planets' steady light: no matter how tiny they seem to us, they do have finite disks that are entire collections of closely spaced points. Of course, from a platform in space outside the terrestrial atmosphere, even distant stars shine without twinkling so our discriminant applies strictly to the earthbound observer. You can, therefore, appreciate what an impact the Space Telescope, to be launched in the late 1980s, will have on astronomy. From above our atmosphere, it will be sensitive to ultraviolet radiation that doesn't reach the earth's surface, and it will be the first major optical telescope to achieve its theoretical design specifications, freed from the blurry limitations of an atmosphere.

2.2. So what is a star?

Now that we know what is not meant by a star, we can again ask our original question. True, planets shine but they are merely mirrors, reflecting sunlight back to us. The apparent brilliance of Venus or Jupiter is due to a combination of their size, distance from earth, and "albedo" - or reflecting ability. Were the sun's light to be extinguished abruptly, even Venus and Jupiter would be turned off (although not simultaneously since the roundtrip travel times for light from the sun to Jupiter and back to the earth and sun-Venus-earth are not of identical length). A star, then, is defined to be a self-luminous entity, where we place no restrictions on either the nature of the physical processes that create the radiant energy nor on the wavelengths of radiation that characterise its output spectrum. It is appropriate both to sharpen this definition and to enable it to encompass several different situations that are encountered during the early life of a star.

However, there is one major generalisation about stellar life that puts star formation into crisp focus: to be a star is to be involved in a perpetual balancing act. Speaking literally, a star is a giant ball of gas and its existence depends critically upon the interplay of two forces: the sheer pressure exerted on the inside by the staggering mass of overlying layers of gas; and whatever internal forces the star can muster to support this crushing pressure. The stellar tightrope is a battle between gravity, pulling inward, and some kind of pressure at its centre, pushing outward. Gravity is a relentless opponent that never relaxes its grip. The slightest weakening of internal support will inevitably lead to a contraction of the star - a squeezing together of all of its gas into a more tightly bound (see Chapter 2.3 below), smaller configuration.

2.3.   The many faces of energy

Why should this contraction occur? Why shouldn't the response of the star to being squeezed by gravity be to expand instead of contract? Why does a ball bounce down the stairs if dropped, never up? As we shall find, nature is energy conscious; always logical, never wasteful. Physical systems prefer to be in their lowest states of energy: they are lazy just like some of us. The ball heads downstairs because that will involve it in a search for a minimum level of energy. For it to head upstairs spontaneously is not impossible but is exceedingly unlikely, although you can very readily see that, with the intervention of your foot, it could be helped on its upward journey. However, your intervention involved an input of energy to the ball which it didn't originally have. It will be valuable at this point to think about the conservation of energy and to recognise that energy comes in many forms that are interrelated.

Let's use an analogy. We have a man and a ladder and a floor (Fig. 2.3). As the man ascends the ladder, rung by rung, it requires an expenditure of his energy. One type of energy available to him is chemical, stored in the connections between atoms in his body, that can be called upon to perform work external to his body.

Energy that is stored but is available to execute work is designated as "potential" energy because it's potentially available if needed. Each upward rung on the ladder represents a conversion of the man's (chemically stored) potential energy into "gravitational potential energy". It's called "gravitational" because if he stepped off the ladder, gravity would pull him down to the floor. "Potential" because his height could be usefully employed: he could jump off the ladder and catapult a colleague into the air from a seesaw, or leap off a tree limb and knock a bad guy off his horse as he passed below. This stationary position at some height on the ladder would have been converted into motion towards the ground if he did jump (Fig. 2.3b). Motion connotes energy too, called "kinetic" energy, and any moving object has it. It goes up with the mass of the object and with the square of its velocity (the product of the velocity with itself). That's why an encounter with a slowly-moving bus and with a rapidly-moving cricket ball both hurt so much.

Consequently we can understand that the ball, preferring to be in its lowest energy state (relaxed, just as we prefer to be) converted its potential energy at the top of the stairs, into kinetic energy as it bounced downhill. (If you're wondering what chemical energy a ball has available to it, since you never saw one eating energy-carrying foods, it's the elasticity of the rubber that provides its energy reservoir.) Its final destination will be at rest (zero kinetic energy) at the foot of the stairs (lowest accessible potential energy

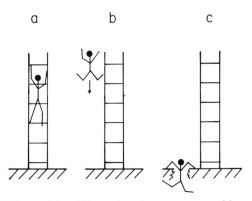

Figure 2.3   Three situations represented by a man and a ladder to show the interchangeability of potential and kinetic energy: a) potential energy; b) potential is converted to kinetic energy; c) a bound situation with negative potential energy.

level), for a total energy (the sum of all forms of energy, potential and kinetic) of zero. "Zero" of potential energy is arbitrarily defined here for there may be a landing or floor below this one that could define an alternative zero if you lived one storey down. In fact, energy conservation does not mean a total energy of zero; merely that whatever value of total energy was first available will always be available.

Let us return to our ladder-climber. Suppose he is a masochist and decides to hurl himself off the top rung of the ladder straight at the floor, combining his internal chemical potential energy (to throw himself off the ladder) and his substantial gravitational potential energy (his initial height above the ground) into kinetic energy of rapid motion. So successful is he in his effort that his feet break through the rather flimsy floor boards and he is left there, suspended at rest, trapped within this floor (Fig. 2.3c). Where did all his energy go?

Did you hear his feet strike the floor? Sound too is a form of energy, communicated by the jostlings of air molecules whose vibrations carry news of the impact to your ears. This was generated, although not electromagnetically, by his collision with the ground. Too, you heard a crash as the wooden subfloor bent and finally cracked, liberating the potential energy stored in its tortured state. Finally, at the moment of impact, our suicidal subject forgot to bend his knees to absorb the force of impact (which would have converted some kinetic energy of his daredevil flight into elastic energy of his joints). The jolt abruptly pushed various leg bones hard against knee and hip joints, leaving them quivering internally, heated up very slightly by the stress and strain of deformation (which was not too elastic as we can tell from the expression on his face!). This is thermal energy. So kinetic and gravitational and chemical potential energies were transmuted into sound and heat. We have forgotten one detail. The poor fellow is still trapped at his waist in the floor and we shall have to extricate him from his predicament - we can describe him as "bound" to the floor.

Think about the word "bound" for a moment. If one is in bonds one is fettered to something substantial. To break free would require the use of a saw, a hammer and maybe a winch; that is, somebody's effort would be necessary to liberate you. To be bound is therefore to have negative energy relative to some zero - to require the input of some positive energy merely to be free again even with a net total energy of zero.

You are all involved in a bound energy situation, though you may not be aware of it, and it's one that involves the constant interchange of potential and kinetic energy. Why doesn't the earth plunge into the sun or go flying off into the depths of space far from its stellar parent? What keeps us orbiting around the sun at our present distance?

It is because the earth is in a bound orbit relative to the sun. It cannot fall into the sun unless some mechanism robs it of energy, removing the potential energy implicit in orbital rotation around the sun. Likewise it can't escape the gravitational clutch of the sun to become an interstellar wanderer, unless it receives from some agency the necessary extra positive energy to achieve this. So here we remain, stuck in our rut around the sun. However, there isn't a constant value of kinetic or potential energy around this orbit. Sometimes the earth is a million miles closer to the sun than its average value; at other times it's a million miles further away than the average. In fact, it travels not in a circle but rather in an ellipse (Fig. 2.4).

At "perihelion" - when the earth is dragged into its closest approach to the sun - it is moving most rapidly. This maximum value of kinetic energy, however, is counterbalanced by the most negative value of gravitational potential energy, since that is when it would require the application of the greatest amount of effort to wrench the earth from its proximity to the sun. Conversely, at "aphelion" - when the earth achieves its greatest distance from the sun - it moves most slowly but is most weakly bound

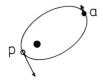

Figure **2.4**   Bound orbit of the earth (small dot) around the sun (larger dot) illustrating "a", aphelion, when the earth's velocity is lowest and "p", perihelion, when the velocity is greatest.

gravitationally to the sun. Yet the sum of all forms of energy (to the accuracy with which we are concerned) is exactly the same at perihelion and aphelion. The value of this total energy determines the dimensions of our elliptical orbit. Were the earth to have been born with rather less total orbital potential energy it would occupy an orbit closer to the sun, as do Venus and Mercury. Greater total orbital energy than our present orbit and we would now be trekking through the coldness of space, at the frigid distances of Mars or Jupiter or even beyond. So, always remember: negative energy means a bound situation. Even remote Pluto, so far from the sun, has a negative total energy to hold it in its place in its curious orbit (that brings it closer to the sun than Neptune for a couple of decades of its 248 year revolutionary period).

Zero total energy in our orbit would permit the earth to escape from the sun although we would not have any energy left to put into motion after our flight. Positive energy and we could adopt the role of interloper - merely chancing to pass through the inner solar system, yet bound to no star at all, in our path that would be influenced by the combined gravitational field of the entire Galaxy.

These are ideas about energy which will be needed in later chapters.

2.4. Stellar collapse

We now return to our fledgling star. Stellar evolution can be thought of as a steady progression in internal density, from the tenuous vastness of a dark cloud of gas and dust, to a final state in which sufficient compression has occurred to raise the internal temperature high enough to ignite nuclear fires (hydrogen will "burn" to helium at some 15 million degrees Kelvin: see Chapter 2.5). This will only happen when the cloud has squeezed itself so tightly that its final density is 10 to the power 20 (that's a one followed by twenty zeroes) times its original! At this point, the compressed ex-cloud could properly be called a mature star since it is supported internally by nuclear fusion (which balances gravity). The main sequence, that we met in Chapter 1, is a locus in the Hertzsprung-Russell diagram defined by the properties of large numbers of stars. It is precisely because stars spend the major portion of their lives in middle age, burning hydrogen to sustain their internal conflict with gravity, that we find so many of them on the main sequence. It is the journey from cloud to the main sequence that we wish to explore in detail in this book, and we will find much to interest us long before the shelter, security or perhaps boredom of the main sequence are attained.

First, we should convince ourselves that the entity we have formed from a fragment of the parent cloud can still be termed a "star" before arriving on the main sequence. We require that it be self-luminous; nothing more. As the initial, very extended cloud steadily shrinks, more and more of its material ends up in a tightly bound, small (compared with the initial state) ball. In other words, we have formed a bound configuration. But the initial cloud fragment was not so bound as will be its progeny - energy therefore has been released. It is precisely because a young star radiates away energy that it is able to continue contracting, a process that may eventually lead to the high density and high temperature necessary for nuclear fusion. So the ball of gas "shines" as it shrinks. It is indeed a star, feeding on lost gravitational potential energy to attain an ever-increasingly bound state.

We use different terms to distinguish the infant from the adult star rather as we do for humans: we have "infant", "baby", "toddler", "child", "adolescent". In the very earliest stages of life, the cloud continues to rain down its matter onto the forming stellar core. During this "accretion phase" we call the infant a "protostar". Once the cloud has ceased to plaster gas onto the core, the accretion phase terminates and the protostar enters its "pre-main-sequence phase", during which it continues to radiate away gravitational energy in favour of core contraction. Once hydrogen-burning is ignited our star has come of age and finds itself relatively close to the main sequence, where the bulk of its life will be spent.

Where on the main sequence does a star wind up? Its endpoint is determined by its starting point: by how much mass accumulated in the young stellar core. High initial core masses imply more luminous, hotter mature states on the main sequence. Lower mass cores evolve into more feebly shining, cooler stars, rather like the sun, to put us in our proper place in the Galaxy.

Theoreticians have not always agreed on the nature of the journey from upper right in the HR diagram to the main sequence. They fall into two rather different camps: "dynamical" modelists and convective-radiative modelists. The first group attempts to follow the hydrodynamical details of a collapsing cloud fragment as it winds through the HR diagram. Published tracks for a few stars are quite convoluted and never accord with observations (c.f. Fig 2.5 and Chapter 5.4). The other camp has been around since about 1950; its more recent contributors have investigated even earlier phases of evolution and differ only in where their calculations indicate that a star joins onto the "convective-radiative" tracks (Fig. 2.5). Let's explain some terminology.

"Convection" is the internal motion in the body of a fluid (liquid or gas) that entails entire chunks of the fluid overturning; dredging up the insides, putting them closer to the outside of the body, then letting them fall back inside again. Your rooms may be warmed by "convection heaters" in which volumes of air enter the device, are heated, rise because they are hot and buoyant, and subsequently

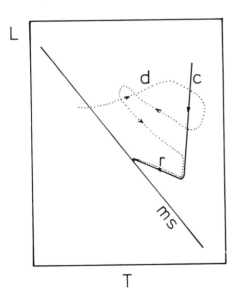

Figure 2.5   Representation of the HR diagram with a dynamical evolutionary track ("d", the dotted path) and a convective-radiative ("c" then "r") track both for a 1 solar mass star. "L" denotes stellar luminosity; "T" the surface temperature; "ms" the main sequence.

fall again as they cool. Stars can be convective in their interiors. This is precisely the mechanism that communicates the energy created in pre-main-sequence stellar cores to the outer layers of these stars. The earliest pre-main-sequence phase, during which gravitational energy is radiated away and heats the core, is represented by a track in the HR diagram called the "convective track" because bubbling convective motions constantly mediate between core and surface of the fledgling star. This extremely rapid (a million years only for a star like our sun in mass) period is followed by one of rather longer duration (around 100 million years) in which radiation provides the link between core-generated energy (still not provided by nuclear fusion) and the stellar surface. We will see these tracks in detail in Chapter 5.

2.5.   Atoms and nuclear fusion

We have mentioned "hydrogen-burning" as the energy source of mature stars. Let's examine this process in more detail. We start with atoms. Atoms are built up of heavy and light pieces. The heavies - protons and neutrons - live in the core of the atom, the "nucleus". They are therefore referred to as "nucleons" and both have about the same mass. Around this nucleus is a cloud of "electrons" - very lightweight particles, weighing in at only about one two-thousandth of the proton's mass. The number of electrons (each having unit negative electrical charge) matches exactly the number of protons (each with unit positive charge), making atoms electrically neutral (neutrons, as their name suggests, carry no charge). Essentially, the mass of any atom is equivalent to the sum of its nucleons' masses.

Hydrogen is the lightest atom in the Universe and consists of a single electron, circling a single proton. The heaviest naturally-occurring atoms on earth are those of uranium, whose elephantine nucleus has a mass 238 times that of the proton. Between hydrogen and uranium there is a host of elements whose atoms have intermediate masses. So where did all these heavy elements come from, given that the Universe is composed dominantly of hydrogen?

Let's try a nuclear thought-experiment. The element helium has a nucleus with two protons and two neutrons, surrounded by four electrons. In terms of its mass, helium is equivalent to four hydrogen atoms (protons and neutrons are essentially interchangeable). So let's try to bring together four hydrogen nuclei, in other words four protons, to construct a helium nucleus. We notice, as we do this, that it's hard work because the four protons don't want to come together. Their electrical charges are all positive and, in electrostatics (the physics of electrical charges), like charges repel one another, while unlike charges (protons and electrons) attract. But we persist and by dint of a good deal of shoving and cajoling we succeed in jamming the four protons very close to one another. We have built a helium nucleus! Just to be sure, we weigh it, fully expecting to find exactly four proton units of mass. And we find 3.97 units. Why? Well, remember all that pushing and shoving? That required a lot of energy from us, to defeat the electrostatic repulsion. That repulsive force seems unusually quiet now - the helium nucleus is very stable - it won't spontaneously fly apart into its constituents. What's holding it together? There is an extremely short-range (it acts over only a few hundredths of a billionth of an inch [a decimal point followed by thirteen zeroes and a 4]) nuclear force, called the "strong interaction", that binds together these nucleons. This nuclear glue produces a bound entity – the helium nucleus. That "binding energy" does the magic for us - it denotes the amount of energy necessary to break apart the pieces of the helium nucleus, and it's related to the energy we invested into pushing the hydrogen nuclei together.

Now Einstein worked to show us the equivalence of mass and energy. His famous equation "E equals m times c-squared" tells us that a little bit of matter (that's the "m") yields a prodigious amount of energy, for the "c" represents the speed of light, already a big number, so its square is enormous in anybody's units. What does this equation have to do with the helium nucleus? The binding energy, which is of course negative, holds together the nuclear components, and can be translated into an equivalent mass - also negative - of minus 0.03 proton masses. This is our mass thief: add together the four protons and the negative binding energy (expressed as this negative mass) and you have exactly 3.97 proton masses, as we observed for our homemade helium. This procedure, of fusing together lighter elements (hydrogen) to make heavier ones (helium) is called "nuclear fusion". But why is this a form of energy? Because the 0.03 proton masses have disappeared in the guise of energy that is lost by the helium nucleus to achieve its fusion. And a tiny amount of mass goes a long, long way in terms of equivalent energy. This is the process that keeps the sun shining, and has done so for the past five thousand million years. Fusion reactors represent the ultimate in the peaceful use of nuclear power since they would yield a far greater efficiency, in terms of energy output per unit mass of fuel input, than any other manmade energy-generating process like burning coal or oil, or utilising wind or tide power.

Does our desire to build fusion reactors therefore imply that these reactions are truly efficient? Let's see. The efficiency for the conversion of hydrogen into helium can be calculated as the energy output (0.03 proton masses' equivalent of energy) divided by the input (4 protons), or 0.7%. So fusion, in an absolute sense, is highly inefficient! But everything else we have is still more so!

This is a good time to appreciate just how massive the sun really is. To shine as it does requires the conversion into helium of four million tons of hydrogen every second! That means the sun has already radiated away energy equivalent to 6 followed by 23 zeroes tons over its five thousand million years of life! Yet, so prodigious is the sun's mass by everyday standards, that, in the absence of other physical processes, it could shine steadily at this rate for 17 million million years! And remember, the sun is only a small star in terms of the Galaxy's overall population of stars.

2.6. Conditions necessary for nuclear fusion

Let's pick up a point that we glossed over briefly. We saw that the hydrogen nucleus is a rather simple entity, consisting solely of a proton. But heavier elements have both neutrons and protons in their nuclei. What trick is involved to convert the four protons of the separate hydrogen nuclei into the two protons and two neutrons of the helium nucleus? How are protons and neutrons transmuted into one another?

Now protons and neutrons don't interact electrically since neutrons have no charge. But they do interact through a force called the "weak interaction". If you thought that the strong nuclear force represented an interaction over a puny distance, wait till you hear the range within which the weak force operates: it's about one hundredth of that of the strong interaction! Yet this ultra-specialised force is the critical intermediary between protons and neutrons. Without it, no heavy elements could be fused from lighter ones in stars. Since the number of neutrons in a heavy nucleus is always roughly the same as the number of protons, the weak force must convert about half the original protons from hydrogen atoms into neutrons to start off the business of fusion.

Thanks to the strong and weak nuclear forces, we see that "all" you have to do, to achieve fusion, is to shove together protons. This is strictly true; however, the shoving necessary is quite out of this world. Well, an alternative to pushing steadily is throwing impulsively. Again, true, but the speed necessary for fusing flying protons is also something not yet achieved on earth. Let's go off on an apparent tangent for a moment. What is "thermal energy"? Heat. What does it mean to say that some fluid, or gas, has a certain temperature? It means that its atoms are moving relative to one another at a certain speed. The hotter a gas, the greater the speed and, therefore, the kinetic energy, of its constituents. Then what temperature would be required to propel protons at one another fast enough to overcome their electrical repulsion? About 15 million degrees Kelvin! So we know that the interior of the solar core must be at least this hot, to achieve hydrogen burning into helium (you see why we call it "burning" even though it is quite unlike any other form of combustion with which we're routinely familiar).

All main sequence stars are burning hydrogen into helium. It is certainly possible for stars to cook up much heavier elements in their interiors, provided that their cores attain sufficiently high temperatures. For example, at a mere 100 million degrees, helium burns to carbon; and at 600 million degrees, carbon to magnesium. Why does it require such enormously larger temperatures to burn these heavier elements than to fuse hydrogen into helium? The reason lies in the greater mass and hence the greater electrical charge of these heavier nuclei. It is necessary to overcome even greater repulsive electrostatic forces when pushing together two carbon nuclei (six positive charges apiece) to make one magnesium nucleus than we encountered during hydrogen fusion. But to achieve such monstrous temperatures demands a stellar core much more massive than the sun's. Even in higher mass stars, that do synthesise the heavier elements by fusion, this is done only in the last stages of stellar life, far beyond life on the main sequence. As such, these processes fall outside the focus of this book. However, we note that the very materials out of which this book, and you, are made were cooked up inside one of these high mass stars!

2.7. The smallest stars

Is there a chance of failure to arrive at the main sequence? The answer is "yes"; if insufficient protostellar mass is compressed into the core, at no time will 15 million degrees be attained by the centre of the star. The hydrogen-burning locus of stellar maturity will elude this ill-fated pre-main-sequence star. This might seem to be something of a freak, not commonly to be met, therefore worthy of dismissal. However, we live very close to at least two such "failed stars", namely Jupiter and Saturn. Technically we should designate these not as giant planets but as "brown dwarfs": stars that haven't achieved and will never achieve hydrogen-burning. Both theory and observations concur in defining a minimum protostellar mass for the ignition of hydrogen fusion to be around 0.08 solar masses. (Jupiter and Saturn, for comparison with this limit, are only 0.001 and 0.0003 solar masses.)

2.8. The biggest stars

Is there such a concept as a "biggest star", a high-mass analogue of the smallest star just defined? Here is a frontier where theory and observation were once in good agreement but may no longer be. Somewhere above 60, but less than about 100, solar masses lie the hottest, bluest stars that we see. A potential cause for the existence of a maximal stellar mass was thought to be "radiation pressure". In the presence of intense radiation, a force is exerted on surrounding material that acts like a wind, tending to blow away this matter by depositing momentum into it (see also Chapter 3.3). It was felt that once a substantial surface temperature had been reached for the protostellar core (some tens of thousands of degrees Kelvin, for example), the sheer pressure implied by all the energetic ultraviolet photons streaming from the core would impede further accretion. Cores would not grow beyond about 100 solar masses. Some astronomers still adhere to this idea. Others, very recently, have felt that a singular ultraviolet-emitting object in one of the Magellanic Clouds (our twin satellite galaxies) may represent a "star" with thousands of times the mass of our sun. In the wake of the idea of such an enormously massive single object, still other astronomers have seized upon several stars in ionised hydrogen regions of a nearby spiral galaxy and upon another, highly peculiar, Galactic star called Eta Carinæ. They have argued that these too may be extremely massive stars. Despite the obvious and topical momentum of this idea there is not yet widespread agreement as to its validity. No-one has any clear view about how to form such massive entities and to defeat radiation pressure in the process. Perhaps it is all a matter of feeding material onto the developing stellar core at such a prodigious rate that radiation pressure is overcome. On the other hand, the conclusion that very massive stars exist is only by inference from the observations.

You know enough already to appreciate the nuances of the interpretation of these observations so let's briefly examine the issues. The principal object is called R136 (Radcliffe 136) and it is the central object in a giant region of "ionised hydrogen" (see Chapter 3.2) in the Large Magellanic Cloud. This galaxy, although a satellite of our own Galaxy, is far, far away. To achieve a sharp, detailed view of R136 is, therefore, difficult. Indeed, the controversy has centred on whether R136 is the core of an entire stellar cluster (and therefore consists of several stars all individually of high luminosity and high mass) or is itself a single super-luminous, supermassive entity. Very recent careful computer analyses, applied to new optical imaging sensors, have been brought to bear on images of R136. These show several spatially-distinct components at the location of "R136", one of which may itself be a very close pair of stars. The new analyses suggest that the critical component is not super-luminous and confirm the likelihood of its being the partially-resolved core of a cluster of hot stars. Now the HR diagram for the sun's vicinity (Fig. 2.6) indicates the relative paucity of hot, high mass stars hereabouts. Increasing our sample to include the brightest stars that we see from the solar neighbourhood adds a few more massive objects but still reveals the dearth of high mass stars. However, the environment of R136 is not even within our Galaxy, let alone in the solar neighbourhood. Therefore, it is perhaps not surprising to learn that most of the stars near R136 are high-mass, high-luminosity objects. So, the "supermassive star" idea is probably wrong, at least when applied to R136, but I hope that you appreciate the controversiality of the topic!

Perhaps both R136 and the bizarre Eta Carinæ are not single entities at all but are entire, compact clusters of very massive, hot (but "allowable") stars. Consequently, for the purposes of definiteness, we shall regard a number close to 100 solar masses as the maximum mass that a single star can achieve.

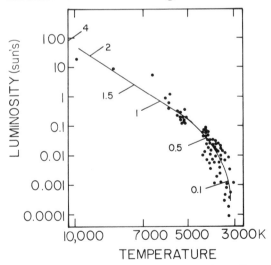

Figure 2.6   Distribution of the sun's stellar neighbours in the HR diagram. Y-axis is in units of solar luminosities. The small figures indicate the masses of stars that lie at different locations along the main sequence represented by the curved line in solar masses.

2.9. Life in the fast lane

One might naively feel that massive stars would take much more time to assemble themselves from parent clouds than stars of lesser mass, like the sun. This is not the case, however. In fact, in the cosmic arena, more massive objects evolve much more rapidly than do lighter ones. Everything about massive stars is fast and brief: their infancy, their duration on the main sequence, their rate of output of energy. Even though they begin life with access to far greater stores of potential nuclear fuel than stars of lower mass, still they are vastly more prodigal with these resources. For example, on the main sequence, a star of 10 solar masses has a luminosity 10,000 times that of the sun, and consequently a lifetime only one-thousandth of the sun's, which amounts to a mere 10 million years on the main sequence (compared to the sun's estimated maturity of 10 thousand million years, of which it has so far used only about half).

2.10.   The stellar mass spectrum

It is instructive to ask, then, just how frequently are these spendthrifts encountered in our Galaxy? Suppose we take a stellar census out to a radius of 20 light years of the sun. In this solar neighbourhood we find a much greater frequency of low mass, very cool stars than of solar-type stars (Fig. 2.6) and none of the massive stars we were just discussing. This kind of distribution, characterising the relative frequencies of stars of different mass, we call a "mass spectrum".

Where did the sun's neighbours originate? First, they are not necessarily contemporaries of the sun: in fact, almost none of them are. A single stellar nursery widely disperses its products throughout the Galaxy over the thousands of millions of years that have been available to it (exactly why this is so is explained in Chapter 3.8). Secondly, the present-day neighbours of the sun represent the results of star formation in many different locations throughout the Galaxy and at many different epochs. In short, the observed mass spectrum must be an average over an entire Galactic ensemble of different nurseries. Does each nursery yield the same mass spectrum, or were the stars of different mass born in different places? This is a most interesting question and one whose answer is amenable to direct observational effort (we shall defer this answer until Chapter 7, however). Incidentally, one subtlety which we have omitted to mention is that one must correct the observed mass spectrum for the effects of differential longevity. In other words, since massive stars last for a shorter time than lighter ones, we expect to see fewer of them, even if stars were created initially with identical frequencies at all masses. Once the observations are corrected for this bias, the resulting intrinsic mass spectrum is termed the "birth function", or the "initial mass function", for obvious reasons. So the question we posed above can be recast in the form: how closely do different nurseries manufacture stars in accordance with the birth function? Of course, if all conformed to this function we would have to find the mechanism responsible for this cosmic collusion, since there cannot be much of a dialogue between nurseries in vastly different parts of the Galaxy.

Armed with these fundamental, global ideas about star formation it is now appropriate to examine in more detail the backdrop for these processes, namely our Galaxy.

3

Our Galaxy

Philologists who chase
A panting syllable through time and space
Start it at home, and hunt it in the dark.
William Cowper
"Retirement"

Stars in their stations set,
And every wandering star.
L. P. Johnson
"By the Statue of King Charles I
at Charing Cross"

---

*Quantum mechanics – ionisation – the hydrogen atom – the
Doppler shift – Galactic structure – spiral arms – density
waves – molecules – dust grains – dark clouds – evolution of
clouds – magnetic fields – polarisation of light – synchrotron
radiation*

3.1. Quantum mechanics

It's time for a digression to the microcosm of the atom
before we tackle the enormity of the Galaxy. We need to be able to
talk about the radiation that atoms and molecules emit and absorb
so we can discuss some of the important astronomical tools that can
be used to infer the physical properties of stars. Quantum
mechanics is a branch of twentieth century physics that grew out of
the embarrassments of classical physics. The picture of the
physical universe that had sufficed for several centuries, since
Newton and his contemporaries invented it, was quite inadequate
to explain the observed behaviour of those tiny entities, atoms.
Quantum physics has successfully replaced its outmoded
predecessor and will give us an explanation of "atomic spectra"
which serve essentially as electromagnetic "fingerprints", unique
to each element.

Quantum mechanics says that light can be thought of in two equivalent ways: either as waves, or as particles called "photons". Photons serve as carriers of bundles of energy – called "quanta". Associated with any particular photon is its quantum of energy and this energy is precisely proportional to the frequency (or colour if it is a visible photon) of the radiation it represents. In this way quantum physics formalises the common idea of "energetic radiation" such as X-rays or gamma-rays: photons at these very high electromagnetic frequencies carry a lot of energy. By contrast, infrared and radio photons are extremely weak because their frequencies are so much lower than even those of visible light. Photons can carry whatever amount of energy with which they were originally endowed. The next concept of quantum physics that we'll meet relates to the very different laws that govern the energy of bound electrons, as opposed to the freely moving photons.

A curious feature of the Universe is the way in which it is composed of an alternation of highly dense and highly vacuous states. Imagine the Universe encapsulated on a microscope slide. You bring this up to your objective lens, roll the focussing wheel and peer into the eyepiece. At first all you see is emptiness: the vast void of intergalactic space that occupies most of the volume of the Universe. Then you encounter a galaxy. You change focus, increase the power and stare at this seemingly solid collection of gas and stars. It turns out to consist very largely of empty space too, between the occasional stars and clouds of gas. Next you scrutinise a star, obviously one of the elemental building blocks of this galaxy. It initially appears solid, so you crank up the power of your remarkable microscope and peer into the "surface" of this star. No longer solid, the gaseous atmosphere of the star is nothing more than a collection of relatively isolated atoms, whizzing about at great speed and occupying an awful lot of otherwise empty space. Undaunted, you turn your gaze on an atom, perhaps expecting to meet, at last, something firm. Here are electrons, circling around a tiny, heavy core – the nucleus of the atom – but between the nucleus and the locations of these lively electrons is empty space again! How can it be that the everyday Universe of seemingly secure and solid macroscopic entities is mostly a void?

One of the fundamental aspects of quantum physics implies that electrons, which are bound to a nucleus, can't choose to move around willy-nilly. They're more like planets, spinning about their suns, in fixed bound orbits. Electrons can occur only in these separate, discrete "shells" and each atom has its own definite set of shell positions. The mere existence of these shells dictates that you don't fall through the armchair in which you're now sitting as you read this. It is the electrical forces between the electrons' shells in the atoms of your body and those of the chair (or between the atoms in your feet and those of the ground) that prevent you from falling right into the earth's core! That's how the world can feel so hard, so unyielding to the touch when you know it is mostly empty space.

Back to the electrons in their bound orbits. Recall from our discussion of planetary orbits (Chapter 2.3) that each such orbit is associated with a specific amount of energy, and that to change orbit necessitates a change of total energy. For example, the earth would have to lose energy to move from its current orbit to that of Mercury, so much closer to the sun and therefore more tightly bound to it (even more negative energy) than the earth. Likewise, to escape to the outer recesses of the known solar system, perhaps to the orbit of Uranus, would necessitate the intervention of a cosmic boot to provide the extra energy that Uranus has, being less tightly bound to the sun than is the earth.

Believe it or not, this analogy provides a clear picture of the atomic spectra, through quantum mechanics. Instead of planets, think of electrons in their discrete and non-intersecting orbits. For the sun, substitute the atomic nucleus as the binder of electrons. Pretend now that you are an electron, trapped in some specific orbit that you don't like. What are your options? If you want to drop into another orbit, closer to the nucleus, then you can, but only if you pay the exact price and give up the difference in energy between your present orbit and the new one. How do you engineer this payment? You pay through the brokerage service of a photon; that is, the extra energy that you want to give away is "emitted" in the guise of a photon of precisely the correct frequency to carry off this much energy. It doesn't matter which orbit you began with, nor which closer orbit you wanted to attain – just emit an appropriate photon and you can take up a position in your desired, more tightly bound, orbit or "energy level". These changes of electron orbits are called "transitions" (Fig. 3.1) and every conceivable transition between a more and a less energetic level (orbit) radiates an escaping photon of predictable frequency. No two elements have precisely the same transitional energies so the spectrum of emitted photons is uniquely diagnostic of the specific atom involved. These atomic spectra are indeed fingerprints, sufficient to incriminate a specific element.

Suppose you made a sad mistake: life close to the nucleus was not all you had thought it would be. You want to go back to your old energy level. How do you achieve this reversed transition? If you're lucky you'll find that photons occasionally pass your atom. However, any old photon won't do. If you try to seize a photon with not quite enough energy to make the jump up to your old level, you'll find that you are simply not permitted even to make the attempt. Quantum mechanics permits only a specific set of energy levels for your atom and "not-quite-an-energy-level" is no level at

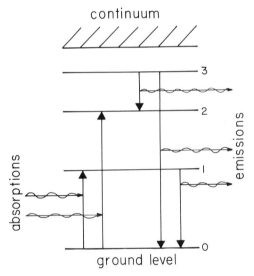

Figure **3.1** Bound electron energy levels in an atom are shown by numbers: 0 represents the ground (lowest) energy state with the most negative energy; 1,2,3 represent levels with increasing energies until the continuum of free electrons is attained (all energies above the zero-energy threshold). Shown are absorptions of photons that drive electrons from the ground to levels 1 and 2, and emissions of photons that correspond to downward electronic transitions from levels 3 to 2, 3 to 0 and 1 to 0.

all. Likewise you will find, unjust though this may seem, that even were you to get hold of a photon with too much energy for the transition, the jump is still not permitted. You have to gauge your cooperating photon's energy exactly. When one eventually comes your way, you grab your chance and "absorb" the photon (the opposite of "emit"), winding up in your desired higher energy level, whilst the world outside has lost a photon to your atom. Conservation of energy is clearly involved but it's as if the currency of energy exchange had a very precisely minted set of coins. Only transactions that involve trading back and forth in the amounts of these coins are legal.

Since absorptions occur at the same frequencies as do emissions, atoms are uniquely identifiable by either their emitted spectra, when they radiate away photons, or their absorption spectra, when they remove these photons from whatever background radiation happens to be available to them.

Let's return to what you could describe as an unjust situation, were you an electron with emotions – namely the attempt to make a transition to a higher orbit using too energetic a photon. Suppose you seized an extremely powerful photon – perhaps a passing X-ray photon. Now you have more than enough energy to jump to any higher bound level – in fact, you have so much energy that you could become a free electron, no longer obliged to orbit your old nucleus. Were you to opt for this exciting prospect then you would become free, flying away, with whatever kinetic energy this transaction leaves you, after paying the exact energy fee for escape from your original level. Your old atom has lost an electron and is now termed an "ion", and the X-ray photon is said to have "photo-ionised" the atom. Had the photon been a little less, or more, powerful you would still have escaped but your speed (kinetic energy) when free would have been correspondingly less, or more, than it actually turned out to be.

Incidentally, since hydrogen has only one electron, it becomes "fully ionised" once this solitary electron escapes. This is not true of heavier atoms. Take helium, for instance, with its four electrons. You could rob it of anywhere from one to all four of its electrons, each time leaving a different "ionisation state", each with its own unique pattern of permitted energy levels. Astronomers distinguish the different ionisation states of an atom by Roman numerals with I denoting the neutral atom (full complement of electrons), II the singly ionised state (one electron lost) and so on. Of course it takes a special environment to yield multiple ionisation of an atom because the removal of more and more tightly bound electrons (the inner ones) requires higher and higher energy. In fact, it takes a very hot environment rich in energetic ultraviolet and X-ray photons to winkle out the final electrons from an atom. In the exceedingly hot and highly rarefied outer layers of the sun is the corona, where the gas temperature attains two million degrees absolute. Here one can see photons from calcium XV, for example, which represents the calcium atom, stripped of fourteen of its twenty electrons!

There is a reverse side to this situation too. You're a free electron, wandering through space, minding your own business, when you come upon an ion. This ion captures you, into one of its more distant levels, in return for which culpable behaviour it emits an appropriately energetic photon. It is then said to be in an "excited state". So may you be! But, just as a ball heads for ground, so this atom heads for its lowest energy state, called its "ground level", in which you wind up as close as possible to the nucleus. To get there you too may simulate the bouncing of a ball downstairs; that is, you can jump from level to level, getting ever closer to the lowest state, and emitting photons as you go. An outside observer, viewing the spectral emission of these photons of different frequencies from the atom, would know of your fate and would describe the process of your capture as "recombination", and the pattern of several predictable frequencies of photon as "recombination radiation" resulting from your "cascade" to the ground level.

3.2. The hydrogen atom

It was the advent of radio astronomy and a prediction in 1945 about the physics of the hydrogen atom that first unveiled a global pattern within our Galaxy. The neutral hydrogen atom ("HI" for "H-one": a single electron orbiting a single proton), when at rest, emits radiation at a wavelength of 21 cm. You may be wondering what transition between hydrogenic energy levels this frequency represents. It's not, in fact, a simple electron level transition. The ground orbital state of hydrogen is actually split into two levels whose energies differ by a microscopic amount. This hairsplitting is called "hyperfine structure", which is due to another quantum mechanical effect. For a moment, imagine the proton of a hydrogen nucleus and its orbiting electron to be a couple of sensibly-sized bodies that you can look at in some detail. Just like the sun and the earth, which spin on their axes, so do electrons and protons. But in which direction do they spin? There are only two choices – clockwise and counter-clockwise. Of course, each proton and electron can choose, but there are only two different configurations. In one, both spin in the same direction ("parallel spins"); in the second, the electron's spin is opposite to that of the proton ("anti-parallel spins"). Quantum mechanics predicts that the parallel spin state has a mite more energy than the anti-parallel state. Further, hydrogen atoms are apt spontaneously (without the intervention of any external agency) to flip from the parallel spin state to the lower energy anti-parallel state. To do so involves a release of energy which is carried off again by a photon but, so small is this energy difference that the photon has an extremely low frequency and an equivalently long wavelength (21 cm). In case you're wondering how frequently this "spin-flip" transition occurs, it happens only once every 12 million years! This is a phenomenon that is not observable in the laboratory for astronomers just can't wait and watch long enough. However, space is so vast and any line of sight through it so long, that even this minuscule probability of occurrence is overwhelmed by the sheer numbers of hydrogen atoms available to undergo the spin-flip. Twenty-one centimetres corresponds to a frequency accessible to radio techniques and one that travels relatively untroubled through interstellar space.

Suddenly the vast volumes of space were no longer silent. The voice of hydrogen – the most abundant element in the cosmos – was audible. Both the spatial distribution and the velocity along any line of sight could be determined; the former from the brightness of a patch of sky at the hydrogen frequency, the latter from the Doppler Effect.

You may have heard of "Range Doppler" in connection with radar. Imagine (Fig. 3.2) a star, emitting radiation in your direction, but also moving relative to you. If the star approaches you, it squashes the waves into a shorter distance than would be the case if it were stationary. Yet the same number of wave cycles is being emitted so this compression is tantamount to diminishing the wavelength, since the same wave train must now fit into a reduced distance. In other words, the wavelength shrinks, the frequency increases. We call this effect the "Doppler shift"; this particular example would correspond to a "blue shift" (blue has a higher frequency than red light). Conversely, were the star to recede from you, the same number of wave cycles now would have to expand to fill the increased distance from you to the star, leading to an increased wavelength, a decreased frequency, and a "red shift" of the radiation. All we need to know is the precise wavelength emitted by an atom, or molecule, in the laboratory (at rest relative to our equipment), and the actual wavelength or frequency received from the celestial source. The amount of blue or red shift

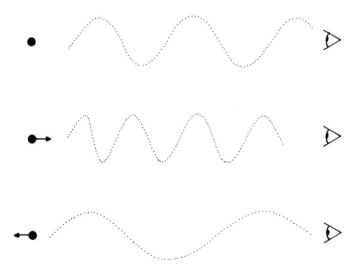

Figure **3.2**   The Doppler shift: a star with no motion along our line of sight does not affect the wavelength of a specific atomic transition. However, a star approaching us squashes the waves into a smaller distance (blue shift) while one receding from us expands the wavelength (red shift). (All effects are greatly exaggerated.)

translates directly into the velocity of approach or recession, respectively, along the line of sight to the object.

By this means, hydrogen maps reveal the dynamics of gas in our Galaxy. They present fascinating snapshots of the "unseen Galaxy" (Fig. 3.3). The great merit of neutral hydrogen is that it shows us the bulk of the interstellar medium, which is relatively cold. We have other probes of conditions in the ionised gaseous zones ("HII regions" for "H-two regions") that surround hot, massive stars, but the birth function tells us that such stars are relatively scarce and, worse, their surrounding ionised regions occupy only a minute fraction of interstellar space. The plane of our Galaxy is very obvious – this is where most of the gas and stars reside. But we also see enormous streams, loops, spurs, bubbles, and magnetically-orchestrated arches that perturb the flattened disk, signifying past explosions and the blowing away of gas by groups of hot, massive stars in the disk. These bubbles are blown by hot stars for two reasons. We have already mentioned radiation pressure in Chapter 2.8 – the steady outpouring of ultraviolet photons that couples momentum drawn from the star into the surrounding gas and especially into the dust. As the dust grains accelerate away from the sources of all this luminosity, they collide briskly with the gas in which they are embedded. They literally drag the gas along with them. There are certainly more direct

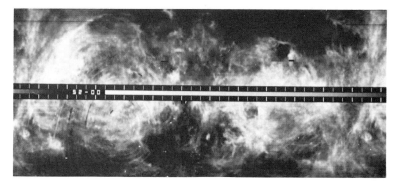

Figure **3.3**   Our Galaxy viewed in neutral hydrogen. Shown is material above and below the Galactic plane – the plane itself is excluded to highlight the intricate structures that arch above and below it. The picture refers only to hydrogen atoms with close to zero relative velocity (1 km/sec blue shift to 1 km/sec red shift) along the line of sight. (Courtesy C. Heiles, University of California, Berkeley.)

methods of coupling starlight to gas but these need not concern us for the moment. It is yet another, albeit inelegant, mechanism that we can identify as a contributor to bubbles in the Galactic gas distribution. This mechanism relies on the existence of strong stellar winds emanating from hot stars; veritable stellar gales, a

Figure 3.4   A pretty face-on spiral galaxy, NGC 2997, illustrating central bulge; a pair of spiral arms; dark dust lanes near the centre; ionised hydrogen regions and clusters of stars strung like beads on a string along the outer spiral arms. (Photograph kindly provided by David Malin, Anglo-Australian Observatory, with the permission of the Director.)

million, and even a thousand million, times more vigorous than our solar wind. Such strong winds deposit a tremendous momentum into the surrounding gas which is slowly accelerated from rest to some low velocity. In brief, we have produced a stellar snow plough whereby a high-velocity but low-mass outflow from a star manages to sweep up a much larger mass of interstellar gas to a far lower velocity. (This represents the conservation of momentum, spelled out in more detail for directed winds from young stars in Chapter 6.6.)

3.3.   Galactic structure: the disk

While I was gaining my Ph.D. at Cambridge University, my thesis advisor introduced me to his concept of "right arm problems". By this he meant there were some questions for whose solution an astronomer might be willing to relinquish his right arm. My advisor had one such issue whose resolution he would really have loved to know: what does our Galaxy look like from outside?

Lest this seem a trivial question, let us remember that it is easy to photograph galaxies external to our own Milky Way system and, from these photographs, to interpret the structure of these gigantic stellar systems (Fig. 3.4). To our perspective, most galaxies reveal the secrets of their morphology, if not of their basic dynamics too. There is little in interstellar space to obscure them from our gaze. However, the structure of our own Galaxy must be intuited indirectly, albeit from a wide variety of observations. We cannot see the wood for the trees.

Even a simplistic sketch of a galaxy such as our own is revealing (Fig. 3.5). Atomic hydrogen maps show us that we live in a spiral galaxy which consists dominantly of a gigantic flattened disk, a central bulge, and a much sparser halo. Its characteristic diameter is about 100,000 LY (or 30 kpc) and we are by no means the largest galaxy for which we can estimate the dimensions. The sun lies about two-thirds of the way out from the centre of this Galaxy and quite close to the midplane of its disk. When astronomers first began to grapple with the details of our Galaxy's construction, in the early decades of this century, they used solely visual techniques. A study of the distribution in our sky of external galaxies indicated that these systems populate, not the entire sky, but most of it. Yet there is a sizeable region in which we rarely, if ever, observe external galaxies. This area was nicknamed the "forbidden zone" and its recognition materially assisted our understanding of the kind of galaxy in which we find ourselves.

As Fig. 3.5 shows, we have an unobstructed view of most of the sky from our location but, towards the very centre of our Galaxy and in the opposite (anticentre) direction, we encounter the disk itself. The disk is not merely a near-vacuum sprinkled with myriads of stars – it also contains vast volumes of gas (mostly hydrogen) and dust. It is the least abundant (by mass and sheer numbers) of the disk's nonstellar components that creates the greatest obstruction along our lines of sight through this huge wheel, namely the dust grains. These particles can be as small as a few millionths of an inch yet one has only to see the dark lane in the sky, which is the Great Rift of the Milky Way through the constellations of Cygnus and Aquila, to appreciate their combined effect on our view of the Galaxy. The dust grains occur on several scales in the Galactic environment: in the largely empty spaces between the stars (the "interstellar medium"); in the immediate

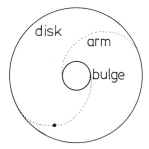

Figure **3.5** Our Galaxy in schematic: top (elevation) shows the disk and central bulge and the sun's approximate location (dot) some 30,000 LY from the centre; bottom (view onto the disk) shows the same ingredients and two spiral arms (dotted) emerging from the bulge.

vicinity of certain classes of star ("circumstellar" shells); and in dense massive dark clouds in which anything from a few hundred to a few million times the mass of our sun is clumped together in the form of gas and dust, and in which, we shall find, new generations of stars are even now being born. These latter enormous structures lie dominantly in the Galactic disk and if we look towards the centre of our Galaxy, we must inevitably look right into several giant cloud complexes.

The Galactic disk is a huge rotating structure but it does not spin like a wheel – like a rigid body – where the periphery always travels faster than the centre. It is in "differential motion". The local speed of rotation depends upon distance from the Galactic centre and is much more rapid closer to the core (beyond an initial small region in which the speed grows as one moves further out). At the solar distance this local rotational speed is 250 km/sec (about one half a million m.p.h.!). Yet so vast is the Galaxy that it still necessitates 200 million years for the sun to orbit once around the Galactic centre!

Imagine what would happen, were you a cloud complex spread over a radial distance of 100 light years or more, at roughly the sun's distance from the Galactic centre. Your inner parts would travel significantly faster than your outer parts and, with the passage of enough time, you would be sheared out in the direction of Galactic rotation, and torn apart. This is certainly one process that limits the life of large, radially extended, structures in the Galaxy.

3.4. Galactic structure: spiral arms

Within the disk itself we perceive pattern and global organisation: a series of what we recognise to be spiral arms loop around us – titanic structures that maintain their integrity for hundreds of thousands of light years. When one is confronted by such enormous structures that are governed by such blatant geometry, it is necessary to understand what force binds the gas together and holds it so rigidly to this cosmic scheme. The weakest of the four known forces of nature is gravity. For example, the ratio between the gravitational and electrostatic forces that are experienced between two isolated electrons is about 10 to the power 38 (that's a one followed by 38 zeroes)! Yet gravity is the supreme example of "action at a distance". Gravity it is, that maintains spiral arms in galaxies (Fig. 3.4).

There is one obvious question about the spiral arms: do they wind ever more tightly around the centres of galaxies or do they unravel? A combination of optical observations (that tell us when obscuring dust lanes lie on the inner edges of hydrogen arms) and atomic hydrogen Doppler-determined velocities (that yield the sense of rotation of a galaxy [does the left side of the galaxy approach and the right side recede, or vice versa?]) leads us to conclude that spiral arms represent a "trailing" system; that is, one that unravels as it circles the nucleus of a galaxy.

But what are the ingredients of a spiral arm that render them obvious to us? First, we have a concentration of neutral hydrogen clouds in these arms. Secondly, if you look carefully at an optical photograph of an arm you will find it consists of individual, extremely bright, blue stars; of whole clusters of these bright stars; and of glowing clumps of gas which emit bright lines of hydrogen, helium, oxygen, sulphur, etc., as a consequence of being bombarded by ultraviolet photons from hot stars, which stars may or may not be directly observable by us. Why should all these hot stars, which are objects much more massive than our sun and therefore have cosmically very brief lifetimes, inhabit the spiral arms? Why are they not equally abundant between the arms – in the so-called "inter-arm" regions? We also find dark, obscuring dust clouds and lanes in the arms: these, we believe, are the sites of present and future star formation. Could star formation be enhanced in the spiral arms? Indeed, are the spiral arms just tracers of high-mass stellar birthplaces?

3.5. Density waves

To answer these fundamental questions we need to take a closer look at the underlying cause and nature of spiral arms. Suppose you are flying in a helicopter over the M1 motorway and below you is an accident in the middle lane. All the traffic slows abruptly to avoid more collisions and a slowing wave propagates back up the motorway. From high in the air you can see a dense knot of cars, moving slowly, disentangling themselves from the actual blockage. This knot continues to move slowly until it is sufficiently far away from the incident, at which time everybody accelerates and the clump dissolves. Individual vehicles enter, travel slowly through, then leave the dense moving clump. This is the nature of what we call a "density wave": it is an instantaneous representation of enhanced density but the individual elements are not locked into the density peak for long. Spiral arms, too, are thought to represent the peaks of density waves imposed upon the rotating Galactic disk. Stars and gas pass into, and eventually through, the arms. The local enhancement in the surface density of the disk is small, perhaps only a few per cent, yet this tiny extra compression of gas and dust clouds that are involved in the spiral arms is sufficient to encourage the clouds to collapse and to make new stars. You shouldn't start thinking that there's a sizeable excess of mass in the spiral arms because that's quite inconsistent with the tiny amplitude of the enhanced density wave. It's just the brilliance of the stars in the arms that renders the spiral pattern so obvious in photographs. The bright blue stars in the arms are a conspicuous clue to the fact that many high-mass (more than ten solar masses) stars have been born of this compression. Having such short lives, these bright beacons of recent star formation never move far from their birthplaces before they die, perhaps spectacularly, then fade away from our visible ken. Low-mass stars, too, could be created in the arms but their much tinier luminosities – even collectively – are too small for them to be recognised as a major visual component of the population of spiral arms. Colour photographs of spiral galaxies show this very clearly: the arms are bluish compared with the yellowish galactic bulge (where a lower-mass but highly abundant population preponderates).

The origin of spiral density waves lies in tiny perturbations that pre-exist in the rotating pancakes of galactic disks. Disks are never precisely uniform and even very slight gravitational perturbations due to small density inhomogeneities will grow and develop into a self-sustaining system of spirals (Fig. 3.6). It is from the differential rotation of galaxies that these waves draw their energy.

In fact, these waves act as transporters of "spin" or "angular momentum" (Chapter 5.6) outward in galaxies.

3.6. Molecules

One great merit to having several, very different, spectral windows through which to view the Universe is that, by this means, astronomers can confirm or reject hypotheses. We have discussed the atomic hydrogen images of our Galaxy and optical pictures. There are two other basic views that have contributed greatly to our understanding of the Galaxy and, in particular, of those putative nurseries of stars – the dust clouds. One of these is the molecular perspective: the other the dust. Not only atoms radiate at specific sets of frequencies that are unique, like human fingerprints; molecules – ensembles of two or more atoms – also are governed by quantum mechanics.

An important set of processes that occurs in dark clouds is the production (and sometimes destruction) of molecules. In such clouds, where the temperatures are very low (only ten or twenty degrees above absolute zero) and the gas densities are quite high (especially when compared with the void of interstellar space), the

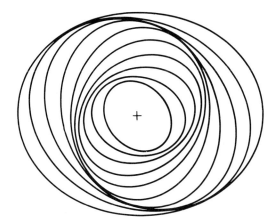

Figure **3.6** The result of gravitationally perturbing stars in elliptical orbits in a galaxy, whose initial ellipses were originally all aligned. Note the "spiral density wave" whose structure results from the proximities of the now misaligned ellipses. (Photograph courtesy of Frank Shu, University of California, Berkeley.)

surfaces of dust grains provide both an environment in which atoms may come together and join into molecules, and shielding from potentially destructive ultraviolet photons that career through space, far from the stars which originally produced them, and nibble at the edges even of dark clouds. Ultraviolet photons carry relatively high energies that could be used to "dissociate" (break up) molecules. You see, atoms can be bound together into the larger building blocks of molecules but this binding is vulnerable to the arrival of sufficient energy to liberate the atoms from their "potential well", before they are off again, freed of any association with their atomic neighbours. What are these molecules that we expect to find, built up in the clouds, and what evidence do we have for their existence?

Molecules must respond to the cosmic abundances of elements. There is little point in our considering the properties of some exotic molecule, for example hafnium bicarbide, when the Universe contains only very limited quantities of hafnium! Cosmic abundances tell us that there is a vast reservoir of hydrogen atoms to be used as basic building blocks for molecular construction, with appreciable quantities of helium, carbon, oxygen, nitrogen, and lesser amounts of neon, silicon, magnesium, iron and sulphur. All other elements are found only in microscopic traces. However, even this simple chemistry set is sufficient to assemble a host of organic and inorganic molecules (with, and without, carbon, respectively). The simplest molecules would be diatomic; that is, molecules consisting of two atoms only. Of these, we have molecular hydrogen, carbon monoxide, carbon monosulphide, etc. Next come triatomic structures like simple, everyday water (two hydrogens and one oxygen atom), hydrogen cyanide (HCN), and so on, through the gamut of organic grammar school chemistry.

Consider a very simple molecule, like the hydrogen molecule. We represent this by a tiny dumbbell (Fig. 3.7). The two "weights" are the hydrogen atoms; these can stretch along the line joining them, or bend relative to this line, or rotate, by twisting around the axis (of course the axis is not a material entity, merely a theoretical construct to represent the potential force that binds the two atoms into a molecule). Molecules can behave in combinations of these activities too. Each of these "modes of vibration", or combinations thereof, corresponds to changes in different kinds of energy at different frequencies. These frequencies dictate where in the spectrum energy would be seen, were the molecule to radiate in these modes. For example, if the two hydrogen atoms rotated about their axis, the emission would be observed in the middle infrared regime, at a wavelength 56 times as long as that of visible (yellow) light. But were the atoms to vibrate by stretching, that would produce near-infrared radiation at only 4 times the wavelength of visible light. A combination of vibration with rotation causes not single frequency lines (as for atoms) but a "band" – an entire ensemble of lines, around the purely vibrational frequencies.

Rotational energies are always smaller than stretching or bending mode energies. Indeed, for many, more complex, molecules, pure rotational transitions (fingerprints) occur only in the microwave region, at wavelengths of a few millimetres or even centimetres. It is the opening up of both infrared and microwave windows that has given us most of our information on the distribution   and   physical   characteristics   (temperatures,

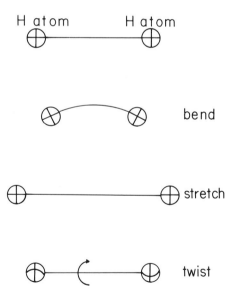

Figure **3.7**   The hydrogen molecule consists of two hydrogen atoms bound together by a force represented by the line joining them. The two atoms can bend, stretch and twist relative to one another. Each such motion represents another form of energy for the molecule.

abundances, velocities) of many gaseous molecules.

The chemistry of interstellar dark clouds is of course of great interest to us since it bears upon our own origins – both that of the earth and that of life itself. Exobiologists are adept at chemical evolution; that is, the synthesis of large, complicated molecules from much more rudimentary building blocks. But their theories are only as good as they are relevant: what were the materials available at the birth of our solar system? We answer that by going one step further back in time and asking about the chemical components of the dark clouds from which the sun and its retinue were fabricated.

We live in fortunate times, at least from the perspective of these questions. The boom in microwave radio astronomy since the early 1970s has led to the recognition of many molecular species, primarily through their purely rotational transitions. Not all molecules are so cooperative, however, and the most abundant of all – the hydrogen molecule (two hydrogen atoms stuck together) – is not accessible to these radio techniques. Nevertheless, the hydrogen molecule too has been detected and its abundance in space investigated quantitatively, through its absorption of the ultraviolet light of hot distant stars, and through its near-infrared transitions (a mixture of vibrational and rotational modes). What do we find for the composition of the interstellar molecular soup? Since hydrogen is the most commonly occurring atom in the cosmos it is no surprise to find that its molecule is also the most abundant. Next come carbon monoxide, water, methanol, hydroxyl (OH), formaldehyde, ammonia, and so on, down a long list of molecules, including monsters with no fewer than 11 atoms! In spite of this rich variety, you should always remember that all molecules are, at best, mere trace constituents of the interstellar medium compared with molecular hydrogen. Cosmic abundances set stringent limits on how frequently molecules that contain much more exotic atoms can occur.

Not every molecular cloud shows precisely the same constituents, nor are the abundances of the molecules, relative to hydrogen, the same from cloud to cloud. The guises that these molecules adopt are not all the same, either. For example, molecular hydrogen and carbon monoxide are found predominantly as gases, freely floating in the dark clouds, occasionally meeting and sticking to a dust grain. One component of dark cloud molecules expected to be plentiful is the family of ices. Now you may think that there's only one ice and that's what you find in your refrigerator. In fact, you can combine this with small amounts of contaminants, such as carbon monoxide, methane, or ammonia, to make more complex icy molecules. What's important about these forms of ice is that all are predicted to be frequent in dark clouds, probably in the form of icy mantles on cold dust grains. On these grain surfaces, chemical reactions can occur and it is by this means that ices are constructed. Too, other molecules may be locked up almost entirely into dust grains.

There do remain controversial issues, of course, and you may be surprised to learn that there is dispute over even such a relatively common (and important!) element as carbon. Perhaps as much as 20-25% of all the available interstellar carbon is tied up in the form of carbon monoxide. Neutral atomic carbon gas may account for a comparable amount, with grains of graphite (and to a lesser extent silicon carbide) or less highly-structured carbon accounting for an appreciable (though as yet indeterminate) quantity. Simple hydrocarbons will hold some carbon, as will more complex substances frozen onto dust grains as icy mantles. But the relative dispositions of these different forms – even the very existence of some – are not certain. Clearly we are not yet ready to pontificate on the origins of carbon-based (or any other) life!

An adequate description of the molecular content of clouds must address not merely formation processes but also destruction mechanisms. For example, the sprinkling of hot stars through space leads to a general interstellar field of ultraviolet photons. As you would expect, these energetic photons are readily capable of imparting so much energy to one particular atom (or larger piece) of a molecule that this atom (or simpler molecule) can be broken off; that is, the energy that binds it to the rest of the atoms, captive in the molecular lattice, can be provided by the photon: the atom is liberated; the molecule is dissociated. One can imagine an extreme version of this process in which all the atoms of a (simple) molecule are liberated leading to "evaporation" of the molecule. Indeed, a very modern topic of growing relevance is that of "photo-dissociation regions". These can be thought of as the surface layers of dark clouds, that are vulnerable to attack and penetration by interstellar ultraviolet photons. This kind of an environment leads to an interesting transition between the dense, fully molecular, cores of clouds and the exceedingly rarefied intercloud medium. In this interface, even the grains are not immune to bombardment by ultraviolet photons that can often liberate fast-moving electrons from the grains' atoms. This phenomenon, a description of which earned Einstein his Nobel prize in 1905, is termed the "photoelectric effect", for transparent reasons. Now, if you're a gas molecule or atom, what do these photoelectrons mean to you? Encounters with them ("collisions") increase your speed which causes a raising of the gas temperature (Chapter 2.6). The gas wants to cool, to achieve a restful ground energy state again. Its preferred mechanism for cooling is to radiate away this extra energy in the form of long-wavelength infrared photons. Consequently, a far-infrared photograph of our Galaxy should show lots of bright-rimmed, but otherwise dark, clouds, if taken in the appropriate infrared radiation of neutral oxygen or singly-ionised carbon (the most effective atomic coolants for the gas). Technology has brought us to the point where airborne far-infrared observations can detect these infrared lines from clouds and has stimulated this theory; surely a pleasing cross-fertilisation of theory and observation!

What astronomers want to know about molecules in a cloud is their abundance, or frequency of occurrence, relative to one another and to the hydrogen molecule; their actual number densities (how many molecules per cubic centimetre); and their temperatures. Armed with this information, plausible modelling of the cloud environment is possible, addressing all the interlaced processes by which molecules of different species grow bigger, creating new molecules, or are broken apart, returning to more elementary ones. Incidentally, these number densities for molecular hydrogen are worthy of contemplation. Clearly, the molecular densities in clouds will be less than the densities in circumstellar shells because many of today's cloud molecules were yesterday expelled from the atmospheres of dying stars. A dense cloud, however, can attain prodigious-seeming number densities, of the order of 100 million to 1000 million hydrogen molecules per cubic centimetre. This sounds like a lot, but what does it really mean? One litre of this soup would weigh only about 3 billionths of a gramme! Consequently, even the densest of dark clouds are excellent vacua!

3.7. The molecular ring

The study of Galactic structure, as determined from the distribution of molecules, is well under way, although not yet so far advanced as those earlier studies that depended upon the 21 cm line radiation of atomic hydrogen. Recall (Chapter 3.6) that we don't employ the hydrogen molecule in the radio region, even though it's the most abundant interstellar molecule; rather we have to use rotational transitions of carbon monoxide (CO). The CO gas is the second most numerous interstellar molecule and it gets its excitation (so that it can subsequently radiate this away and be observable) by collisions with hydrogen molecules. Consequently, the spatial distribution of CO is an indirect tracer of the hydrogen and, therefore, of the bulk of the Galactic molecular mass.

The comparison of the distributions of atomic (directly observed) and molecular (inferred from CO) hydrogen with distance from the centre of the Galaxy is fascinating. Within our Galaxy's nucleus (out to a radius of about 5000 LY) there is almost 50 times as much molecular as atomic mass. The real structure inferred from the observations of of the molecular centre is that of an edge-on disk about 250 LY thick and 5000 LY in diameter. Its plane is tilted very slightly with respect to the average Galactic plane (you will find disks to be a recurrent theme of this book, from the very largest to the very smallest spatial scales!). As we move out into the Galaxy, the molecules dip in density but peak up again, 20,000 LY from the centre. Their distribution plummets dramatically beyond 20,000 LY radius whereas the atomic hydrogen is virtually constant. So, not only do we find a molecular core, but even a molecular ring represented by this secondary peak in the distribution. The origin of the molecular core is perhaps more transparent (see Chapter 5.6 on the conservation of angular momentum in rotating systems) than that of the ring, but rotating central disks are both an important topic as well as one currently in vogue.

Within 50,000 LY of the Galactic centre the total mass in the form of molecules is almost equal to the atomic mass although, inside the sun's Galactocentric radius (we live about 30,000 LY out), we find 85% of the mass in molecules but a mere 25% of the atomic material. Therefore we learn about quite different components of the Galaxy by comparing the 21 cm neutral hydrogen and microwave (millimetric) maps of CO. It is inferred that most of the molecular matter resides in many separate clouds (over a wide range of spatial scales) whereas the atomic hydrogen is distributed rather smoothly, is not clumped, and shows relatively little variation with Galactocentric distance. Clouds appear on a wide range of scales; some as small as 3 LY, others as large as 1000 LY. Actually, most of the CO mass resides in giant molecular clouds, perhaps 150 LY across, and holding between 100,000 and a million solar masses of material.

3.8.   The importance of dust grains

Dust plays a vital role in the cycle of stellar birth and death. It enables gas clouds which are collapsing to become dense enough at their centres to create new stars. It collects into disks about these newly formed stars and provides the material for planetary systems. It is produced at the end of life as a sort of smoke that returns the heavy elements, cooked up inside the stars during nuclear burning, to the interstellar medium. Agglomerations of interstellar gas and dust yield new dark clouds and the cycle is complete.

It is worth stressing the relevance of dust to the existence of low-mass stars for a moment. As a blob of gas contracts it gives up gravitational potential energy (Chapter 2.4). But contraction can only be maintained if this unwanted energy can be disposed of. Eventually the centre of the cloud heats up appreciably. As its central temperature climbs, the photons radiated by the core become more and more energetic. But the gas accumulated at the young stellar core becomes opaque to this radiation: it persists in absorbing it instead of letting it escape to the outside of the star. In short, the flow of gravitational energy outward would stop if the star were made purely of gas. However, sprinkled in the gas are dust grains which readily gobble up the energetic photons. The grains warm up too but only to temperatures of a few hundred to a thousand or so degrees, where they radiate infrared photons. These infrared photons can escape through the gas and dust and penetrate to the future stellar surface. The dust grains, therefore, enable the continuation of the process of radiating away gravitational energy; the core collapses again until eventually temperatures appropriate for nuclear fusion are attained. Without dust grains one would attain only a hot-centred gas cloud and not a star. They serve as a valve rather like your finger and thumb, gripping the neck of an inflated balloon: the balloon only collapses if the valve remains open. The situation is somewhat different for the formation of stars

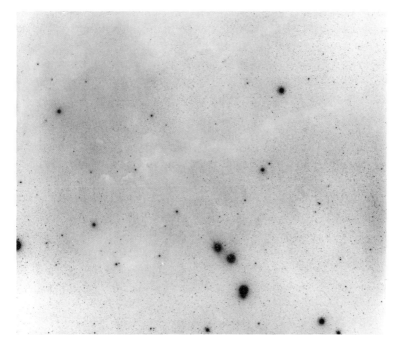

Figure **3.8** A portion of the Taurus-Auriga clouds. Images are in negative so stars appear black, and obscuring dust clouds appear as almost white voids against the stellar background. (Copyright 1960 National Geographic Society-Palomar Sky Survey. Reproduced by permission of the California Institute of Technology.)

of the highest mass but we shall defer discussion of this until later (see also Chapter 9.6).

Made of materials like talcum powder, and perhaps stomach settlers (silicates and perhaps carbonates of calcium and magnesium), mirror-grinding powder (silicon carbide or carborundum) and pencil leads (graphite), these grains pervade space. Although they are widely dispersed in the interstellar medium, dust grains also congregate prodigiously where giant masses of gas come together in the Galactic plane, for dust and gas are well-mixed. These clouds may be recognised on sky photographs (Fig. 3.8) by their ability to block out the light of background stars. These empty whitenesses represent the stellar nurseries – the birthplaces of stars like our sun. We do not see through these clouds: they are simply too dense, too absorbent of starlight. Hence the "forbidden zone" in the sky (Chapter 3.3) represents the plane of our Galactic disk – so densely populated with dusty gas that even details within our own Galaxy are lost to view, let alone the background distribution of distant, external galaxies.

From where do these diverse ingredients of the interstellar medium come? Hydrogen, cosmically the most abundant element, was made early in the history of the Universe. When galactic masses of gas were assembled, soon after the "Big Bang" (the explosion that initiated the formation of the Universe), these were of hydrogen. These masses collapsed to form the wealth and diversity of galaxies that we see today. Such a picture implies that the first generation of stars had to be made almost entirely of hydrogen (and a little helium) and it is thought that these firstborn objects were massive, substantially larger than the most massive stars that we find in our Galaxy today. In the early Universe there were no dust grains incorporated into the pregalactic gas that filled space. It was from the catastrophic demise of the earliest massive stars that heavy elements were churned out into space, the products of thermonuclear cookery in stellar crucibles. From atoms of carbon, nitrogen, oxygen we were all made; from magnesium, silicon and iron came the interstellar grains. Subsequent generations of stars were born of this enriched gas – hydrogen, contaminated by the heavy elements – and they, in turn, gave back some of this material to the interstellar medium during their evolution. Not all of the hydrogen initially available to our Galaxy was incorporated into stars; much still survives as atomic gas in the interstellar medium at large, and as molecular gas in dark clouds.

As we shall now discuss, dust provides a crucial tool for investigating the physical conditions in dark clouds.

3.9. Probing the properties of clouds

The dark clouds are fascinating objects and the determination of their parameters often poses a substantial challenge for the astronomer. One would like to know the following: the total mass of a cloud; the disposition of this material between gas and dust, and between atomic and molecular forms; how much the cloud affects observations of the light of background stars; the temperature of the cloud; its kinematic state – in other words, is it rotating systematically, or stirred up by unseen stars, or expanding, or even more exciting, perhaps contracting and actively trying to make stars? Some of these properties are amenable to direct observation but many depend upon copious assumptions and inferences about the cloud's true state. Let's look into the ways that astronomers can probe the surfaces and the interiors of clouds.

Using only light it is very easy, given a suitable cloud (one that is not too obscuring), to count stars around and through the cloud from a deep photographic plate of the region. By looking at the colours of stars seen through the cloud one can estimate how much redder these appear than their intrinsic colour. To do this with any precision, of course, requires a spectroscopic investigation (see Chapter 5.2) of each star that will yield its probable temperature, hence, true colours. During its passage through the cloud, starlight is extinguished; that is, attenuated, usually in a wavelength-dependent manner. Blue light suffers most; red least. This was discussed in Chapter 2.1, and we note that each extra scattering of an optical photon involves yet another encounter with an atom or dust grain that is only too eager to absorb the photon. So blue light suffers much more than does red. Comparing observed stellar colours with intrinsic ones yields the degree of attenuation of background starlight. The greater the "reddening" of the starlight, the more severe the attenuation. Even if one cannot actually see background stars through a cloud, the observations of diminishing star counts, in different regions that steadily approach the periphery of the cloud, show a trend from which one can roughly approximate the central obscuration right through the core of the cloud. What levels of extinction are actually deduced? Typical clouds permit only 0.01% to 1% of the background starlight to filter through their cores, and these are by no means the thickest clouds. In some extreme cases, a cloud may cause so much obscuration that a minuscule 1 part in 10 to the power of 40 (that is a one followed by forty zeroes!) of starlight incident on the front of the cloud can be detected at the back. Certainly there are many clouds of sufficient obscuring power that, were one suddenly to be interposed between the earth and the sun, we could not detect the sun even from the Space Telescope (due for launch in 1989 and to be our most sensitive optical photometric instrument)! You may wonder what produces all this blocking of light. It is entirely due to the small quantity of dust, mixed into the gas which forms the bulk of the cloud's mass.

A calibration has been established between the number of hydrogen atoms and the amount of extinction along the paths towards a number of stars. Such a relationship is strictly valid only in the interstellar medium which represents a much more rarefied fog than the "pea-soup" of a dark cloud. Nevertheless, with some care, astronomers have applied this calibration even through intervening dark clouds to derive estimates of the amount of hydrogen gas embodied in the individual clouds. What would happpen if you had selected one of the very dark clouds that permitted essentially no starlight to penetrate? How could you estimate the gas content of such a cloud? One technique that extends from the optically thin to optically thicker clouds is to use near-infrared observations. Even an increase in wavelength of only 4 yields photons that can penetrate dust 11 times better than ordinary yellow photons. Of course, one can be unfortunate and find not a single detectable background object even to the limits of sensitive infrared telescopes. In such a situation, the astronomer must resort to very different techniques.

You should never forget that, although these clouds present virtually impenetrable barriers to you, it is not because clouds are solid. Rather they represent exceedingly tenuous media; less dense than a fine terrestrial laboratory vacuum. But the dimensions of clouds are so vast, by our standards, that the cumulative effects of these virtual vacua over light years are all too noticeable.

What about the detectability of the dust grains themselves? At very long wavelengths (in the so-called submillimetre region) of the order of 0.1 to 1 mm, the clouds are exceedingly transparent and every single grain can be seen, no matter how deeply located. This is the ultimate advantage of long-wavelength radiation. Therefore, the thermal emission that we detect from a cloud is a function of the wavelength of observation, and the grains' composition, size, shape, density and optical properties (the latter properties essentially fix the temperature). If we establish the properties of grains then we can model the far-infrared observations and relate the energy radiated at specific wavelengths to the number or, equivalently, the mass of dust particles.

How can we determine the shapes, sizes and nature of this interstellar dirt? Observations of interstellar dust are now routine, all the way from the ultraviolet (using remotely controlled satellites) into the far-infrared. By looking at the absorption properties of the interstellar medium we learn about the character of the intervening dust. All you need is a background source that is bright in the right part of the spectrum. For example, distant hot stars serve as good ultraviolet, optical and infrared probes of the dust in space. Such observations tell us about the likely distribution of grain sizes (there is no reason to assume that all dust grains have the same size), and something about grain composition too. When we look at the radiation we observe from some hot star of known temperature, the difference between what we expect to observe and what we actually see is attributed to the intervening interstellar medium. This kind of experiment tells us grains are present with radii from less than about one half a millionth to ten millionths of an inch, and with compositions that suggest the presence of both graphite and silicates. In the darkest clouds we find that the average size of grains is much larger than that in the general interstellar medium. Why should this be the case? It happens principally because some of the molecules present in gaseous form like to condense on the surfaces of cold grains (just as your warm breath condenses on a cold window pane). This mantle of molecules freezes into various forms of ice (Chapter 3.6) and it is the presence of this frozen overcoat that increases the size of grains within clouds, although the actual mass of an icy mantle is usually less than that of its host dust particle.

It is extremely difficult to infer anything about grain shapes from the observations. Almost all interpretations are carried out under the assumption of spherical particles. However, our ignorance is not such a liability as you might imagine for, unless most of the dust in clouds was in pathological forms such as very long thin needles, or thin flakes, grain shape is quite unimportant in its effects upon our estimates of mass derived from the infrared observations.

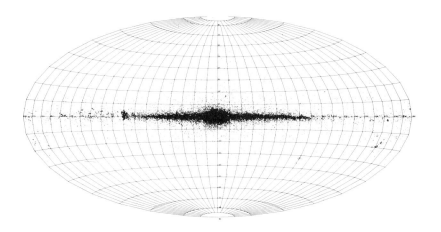

Figure **3.9**   Our Galaxy as seen from within by IRAS. By plotting the Galactic positions of all the objects seen by the satellite, whose shortest wavelength energy and whose temperature lie in carefully selected ranges, one can clearly see the Galactic disk population and the obvious central bulge. Plotted horizontally is Galactic longitude; vertically is Galactic latitude. (Photograph courtesy of Tom Chester, Jet Propulsion Laboratory.)

3.10. An infrared view of the Universe

Infrared can also be used to view simple thermal radiation. The temperature of an object determines at what wavelength most radiation will be emitted. An example of Galactic structure discerned from purely thermal emission by stars and by the dusty components of interstellar clouds is shown in Fig. 3.9. This is a thermal image of the plane and the central bulge of the Milky Way taken by "IRAS" – the highly successful international venture by the U.K., the Netherlands and the U.S., in which the Infrared Astronomical Satellite orbited high above the earth, surveying the entire sky for the first time at long infrared wavelengths and with high sensitivity. Fig. 3.9 is constructed from IRAS's vast database. It portrays the disk as a giant but thin structure, and the Galactic bulge as a very bright mass of thermal infrared emission.

The characteristic temperature of the plane, as seen by IRAS, is between both the typically 10 K molecular and 80 K atomic hydrogen gas and the several thousand degree stellar component. IRAS shows us the cool dust grains, intermingled in gas clouds – both atomic and molecular – that are heated by starlight. This represents an entirely fresh view of the Galaxy; one sensitive to a substantial constituent, never before capable of being studied across the entire sky.

There appear to be both cool (around 100 K) and cold (around 30 K) dust clouds but, so far as we can discern, dust grains are always accompanied by gas and vice versa, although the relative proportions of dust and gas, the amounts of molecular as opposed to atomic gas, even the types of dust grains, vary from place to place.

3.11. Formation and destruction of clouds

Grains play a vital role too in the construction of molecules from atoms. The surfaces of grains act as intermediaries for the meeting of passing atoms, some of which will stick to them. Two, or even several, such atoms may be able to combine into a molecule given the right physical conditions. Subsequently the molecules can leave the surfaces of their host dust grains and travel into the dark clouds which contain an abundance of grains. Why were the atoms and grains originally in such close proximity? The Universe is an essentially empty space; to talk of the void of space is to be precise, yet distances and times are so great that, even in this vacuum, chemistry can occur. Objects at large in the Galaxy do not feel the gravitational fields of their neighbours; rather they sense the combined, average gravitational field of the entire Galaxy and its constituents; yet a close encounter with an appreciable mass will inevitably leave its stamp upon the orbit of the less massive object. So it is that gas clouds and dust grains gather together in a number of deep potential wells, attracted by the already accumulated material. What might create these potential wells – these congregating places for clouds – is still not known although it is believed that the Galactic magnetic field plays a role (see Chapter 7.3). However, once such wells are established, they attract increasing amounts of matter drifting by them. In this manner, and over extremely long periods of time, dark cloud complexes are built; regions of space of the order of 100-300 LY across that contain from 10,000 to one million times the mass of our sun. These are the stellar nurseries that contain the seeds of the next generations of stars in the form of the debris of previous generations.

We see many clouds and cloud complexes in the present-day Galaxy but how long do individual clouds live? Have the present nurseries always existed or does the Galaxy initiate star formation in different regions at different times? What could limit the life of a cloud?

Radiation fields that are very intense act to provide a significant pressure for nearby atoms and grains. If a cloud were to manufacture a very hot star, the radiation pressure of this object would tend to disperse the remaining cloudy medium, blowing an ever-increasing cavity in the cloud which could ultimately puncture it and even cause its disruption. Processes which occur even in low-mass (cool) stars have recently been discovered that jeopardise the continued existence of the ambient cloud in their vicinities too (see Chapter 6). So the very creation of stars, quite apart from using up a finite amount of the total material in a cloud, can threaten the cloud by tending to disperse it.

So far we have offered these mechanisms for the destruction of clouds only as suggestions. In Chapters 4 and 5 we shall consider different observations that bear upon the validity of these possibilities. However, the differential rotation of the Galaxy is always present, in any region, and will always tear apart any radially-extended mass of material: it is merely a matter of time. There is a good deal of controversy among molecular radio astronomers as to how long-lived clouds, that have been well-observed, really are. Estimates vary from 10 million to one thousand million years and both sides are equally vociferous in their opinions. From this we learn that even observations can be equivocal in their meaning. Of course, the Galaxy is a very large volume within which are found small dust clouds and giant molecular complexes: one can readily see that a mechanism like differential shear assumes greater relevance in a geometrically large region than in a small one. Perhaps other selective mechanisms, too, affect clouds.

3.12. Magnetic fields

There is another component to the Galaxy that we must address: an all-pervading yet seemingly directly invisible component – the magnetic field. Like the gravitational field, the magnetic field is very widespread. It threads its way through dark clouds. It can influence the motion of electrically-charged particles and material by exerting a force on those charges, thereby providing some control over clouds that try to collapse. How can we observe magnetic fields? We shall discuss only two methods: one optical, the other radio.

First let us consider the "polarisation of starlight". What do we mean by "polarisation"? Remember that light travels in waves and that it's electromagnetic in nature. That description – "electromagnetic" – means that light consists of an electric field oscillating in one direction, a magnetic field oscillating perpendicular to the electric field, and the wave travels forwards along the third mutually perpendicular direction (Fig. 3.10). So what's special about polarising sunglasses? Well, light waves come in two flavours: those with electric field orientated up-and-down or side-to-side (Fig. 3.10). What polarising glasses do is to allow only one kind of light wave through: they filter out the wrong type of orientation of light wave. That's why they make the sky look darker. The sky consists of blue light from the sun that has encountered the atoms, molecules and particles of our atmosphere and been scattered (Chapter 2.1). This scattered radiation field contains light polarised preferentially in one direction whereas cars, houses, ice cream cones contain both orientations of light wave – they are "randomly polarised". You know, if you've worn them, that these sunglasses also reduce glare from the road when you're driving along it, but only if you're driving the right way relative to the direction of the sun. This happens because the glare on the road is nothing more than reflected sunshine. Every reflection tends to produce polarisation by weakening one or other of the two types of light wave, so reflection creates images of the world that are vulnerable to cancellation by someone wearing their sunglasses in the appropriate manner (Fig. 3.10).

Imagine the light (or infrared radiation) of distant stars traversing the interstellar medium or a dark cloud. The magnetic field in some clouds is strong enough to align electrically-charged dust grains (sometimes grains will be parallel to the field's direction; at other times their orientation can be perpendicular to the field). These preferentially-aligned grains act to select in favour of one or other orientation of electric vector, so they are polarisers of starlight. Therefore, if we measure the direction in which the

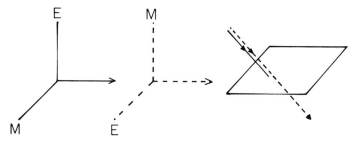

Figure **3.10**   The two polarisations of light. Represented are "E", the electric field; "M", the magnetic field; and the motion of the light wave (arrowed). The right hand figure shows the effect of a polarising filter. Incident are both forms of polarisation (solid and dashed); emergent is only one – the light is then said to be all polarised in one sense (the dashed line).

light of background stars is polarised, we can effectively determine the orientation (and perhaps the strength, subject to a host of assumptions) of the magnetic field in the cloud. On average, the field lies within the Galactic plane although, local to the sun, it has a helical structure, arching almost perpendicular to the local plane.

For the second means of tracing the magnetic field we swing to the other end of the spectrum, to the radio region, and we shall consider a very different kind of mechanism. Space is permeated by large numbers of electrons which, being electrically-charged (negatively, though the sense is not important here), are forced to revolve around magnetic fields. As they do so, they produce radio emission called "synchrotron radiation" which has a very characteristic radio spectrum. When we look at the spiral arms of external galaxies we see a greatly enhanced degree of radio emission from the arms. This has been interpreted as synchrotron radiation that reflects the local compression, and hence increase of magnetic field, produced by a spiral density wave in the arms. So, by this technique, we can map the direction, and even the strength, of magnetic fields. Incidentally, this is another of the important differences in the views of galaxies that radio and optical astronomy give us. Optically we become aware of individually bright stars that are certainly observable at highish radio frequencies too. But, at lower radio frequencies, the galactic disks themselves speak, through the synchrotron emission that is plain wherever the magnetic fields wander and there are electrons to spiral around them.

Now we shall focus upon some real stellar nurseries and discuss how to recognise these; what their stellar and gaseous ingredients are; and what evidence exists for interaction between newly-formed stars and their parent clouds.

4

Where are stars born?

Sky – what a scowl of cloud
Till, near and far,
Ray on ray split the shroud
Splendid, a star!
Robert Browning
"The Two Poets of Croisic"

The jewel of the just,
Shining nowhere but in the dark;
What mysteries do lie beyond thy dust,
Could man outlook that mark!
Henry Vaughan
"They Are All Gone"

Stellar birthplaces – scales of star formation – signposts of star formation – connections with clouds – the "cosmic questionnaire"

4.1. Stellar birthplaces

From our sun to the nearest known star is a distance of 4.25 light years. Space is not crowded yet we should not imagine that the sun was born in isolation. Stars are gregarious, at least in their early lives. They are found forming in crowds, perhaps ten, perhaps a hundred or more. It should be remembered that our eyes, and photographic plates, are attuned to only a minute fraction of the entire spectrum. Dust clouds attend the birth of stars and light is a very poor probe of their inner secrets, unlike radio or infrared radiation. We can observe with light only what we can see. That means we perceive only stars on the surfaces of dark clouds, or slightly below the surfaces, where the total line-of-sight obscuration does not exceed a modest amount, rather than along the path through the entire cloud. This gives us a woefully biased perception of star formation in dark clouds; this must always be kept firmly in mind. However, despite this bias we can observe very young stars – believed to represent the state of our sun when it was only a million years old or less ⊢ in abundance in specific locations. In these regions the young stars are separated from their visible neighbours by anywhere from only one hundredth of a light year to one light year.

If you are wondering where the sun's childhood neighbours and contemporaries have gone, then let's recall the differential rotation of the Galaxy. The sun is now some 5000 million years old. It has completed 25 Galactic orbits since its birth. Even a very slight difference in radial distance from the centre of the Galaxy between two stars, initially adjacent, suffices to separate them by a great gulf after millions of years. For example, if the sun and a childhood companion were separated radially in the Galaxy such that only 1% of the sun's present rotational speed was the velocity difference between them, then, after 25 Galactic orbits, the sun would now be 40,000 light years away from that star, almost one radius of our Galaxy!

4.2. Scales of star formation

We are aware of star formation occurring on several different scales. At the small scales, one can find tiny dark clouds, merely one or two light years across, with only one visible young star. Of course, a thorough infrared survey could well reveal the presence of others, so deeply hidden in the dust associated with the cloud as to be quite invisible to optical telescopes. It is hard to conceive of a cloud creating just a single star but, if we were lucky enough to catch a glimpse of the first-formed star in some cloud, it might well appear that way. Others might follow yet those objects could still be invisible. This solitary star-forming process is known to have made low-mass stars (with about one solar mass) in several clouds. On the next scale, one can find somewhat more massive stars than the sun, recently born in relatively small clouds, perhaps ten light years in diameter. However, these stars are usually found to be accompanied by one or more low-mass visible young objects. Finally, we know of star formation on the grand scale. The nearest nurseries to the sun have manufactured a very high proportion of low-mass stars with either no, or only one or two, higher-mass objects in attendance. In Taurus-Auriga, in Ophiuchus, and in Chamæleon, we have identified entire associations of around 100 visible low-mass stars. These nurseries lie 500 light years away, and are tens of light years, up to even 100 light years, across. When low-mass stars predominate we term the region a "T-association" (why "T" we shall see in Chapter 5). By contrast there exist giant cloud complexes, generally rather distant from the sun (thousands of light years), in which only massive, very hot stars are seen. These stars are termed "O-stars", have masses at least 20 times that of our sun, and live only a few million years (Chapter 2.9) before they evolve into supernovæ or equally dramatic exploding entities. Therefore, these complexes of dust clouds are said to be allied with "O-associations". Again one must be careful to avoid reaching the conclusion that O-associations make solely high-mass stars. Using light alone, we would not have recorded and recognised low-mass stars at the great distances of these molecular cloud complexes, unless special imaging techniques had been used, in combination with the largest telescopes. This introduces another type of bias into our cosmic probing: when celestial surveys are limited by luminosity, the sample of objects discovered is complete only to near distances for faint sources and to much greater distances for more luminous objects. Likewise it is not clear from optical studies alone that the T-associations truly manufacture no, or few, high-mass, for example O-type, stars. But with the additional information of radio and infrared surveys such a statement can be made.

4.3. Giant molecular clouds

We discussed molecular observations of the Galaxy in Chapter 3.7. It is thought that today's Galaxy contains about 6000 giant molecular clouds even in its inner portions: these clouds dominate the interstellar medium. All are bigger than about 60 light years across, weigh in at 100,000 solar masses, and they contain more than 90% of the molecular mass. There are some veritable monsters too – perhaps 500 light years across, containing upwards of a million solar masses! There is real momentum, now, in the idea that most stars born in our Galaxy form in these giant molecular complexes, rather than in atomic hydrogen clouds. This clearly bears upon any discussion of where and how star formation proceeds.

An area of current hot dispute relates to whether CO clouds are confined to spiral arms (hopefully the same arms as delineated by HI!) or are permitted between the arms (in the "interarm" regions). This is a complex issue that necessitates awareness of the pitfall of observational bias in the selection of regions mapped. The cleanest resolution at present might be to attribute high-mass star-forming regions to the inner Galaxy where these regions will map out the spiral arms. Clouds in which high-mass stars are forming are, of course, warmer (by virtue of the heating effects of these hot stars) than cold clouds (not associated with hot, massive stars). Warm clouds are always easier to see, observationally, than colder ones for their CO emission is more prominent when the molecules are at higher temperatures. Low-mass stars on the other hand may form in all clouds and many of these clouds could occur frequently in the interarm regions; but each of these clouds is only a tiny fragment compared with the giant complexes that trace spiral arms.

To corroborate this view that CO clouds map high-mass star formation, we note that radio continuum surveys are highly sensitive to the presence of ionised gas such as that produced in the vicinity of hot stars, by their blast of ultraviolet photons, capable of ionising hydrogen. There is a very strong correlation between the warmest giant CO clouds and giant ionised hydrogen regions. We may not be able visually to see the newly-formed stars because of the thick intervening dust, either purely interstellar (because the cloud is very distant, in the dusty Galactic plane) or intracloud (because the star is deeply embedded in its parent cloud) or both. However, the CO "hot spots" and radio continuum peaks are excellent indicators of the presence of these hot, high-mass stars.

4.4. "Starburst" galaxies

IRAS (Chapter 3.10) has given us our first unbiased, far-infrared look at the Universe at large. It reveals many optically very faint galaxies that are booming infrared sources. What is the cause of all this infrared radiation? For our own Galaxy it has been estimated that 10% of all the bolometric luminosity derives from hot, high-mass blue stars. Likewise, independently, 10% of the total luminosity is radiated at far-infrared wavelengths. We know that giant molecular complexes are correlated with hot stars (seen or unseen, visually). We find that far-infrared maps of the Galaxy also reveal strong peaks coincident with ionised gas regions and with CO peaks. Can we conclude, therefore, that sensibly all far-infrared luminosity ultimately derives from hot, high-mass stars? If so, perhaps the same holds true in external galaxies. In that case, the infrared-richest galaxies, brought to our attention by IRAS, might represent stellar systems in which vast numbers of high-mass stars have recently been born. Most are still heavily embedded in their own especial fragments of parental dark clouds (their "cocoons") leading to the principal characteristic of these galaxies – their prodigious ratios of far-infrared to blue luminosity. Whatever the sponsoring mechanism, it is as if many clouds have been stimulated into bearing (though not yet baring!) stars. Some call these systems "starburst" galaxies since the epoch of star formation must have been very brief.

4.5. Signposts to star formation

Armed with a basic knowledge of astronomy how can one distinguish what is a young star from what is not? One clue comes from the O-stars. Wherever they are found we know that they live for a cosmically brief span; therefore, where they are seen now must be quite close to their birthplaces. Typical velocities for elements of gas within dark clouds are of the order of a few km/sec; therefore, stars born of these pieces of material will also move slowly. Since these O-stars are still visible they have clearly not reached the catastrophic ends of their lives. Consequently, some of them must be only 100,000 to a million years old, during which time they can have travelled no more than 20 light years from their places of birth. At the great distances of typical O-associations this corresponds to an angular separation of less than a small fraction of a degree on the sky. So, to find the nurseries for at least some stars, we need only investigate the immediate environs of any O-stars.

The same argument, of course, applies to isolated O-stars as to entire O-associations. For the intermediate scales of star formation mentioned above, the presence of much fainter stars, apparently in the same clouds from which an O-, or almost as massive, star has formed, also suggests that we should examine these fainter objects carefully in case they too are young stars, albeit of lower mass than the visually dominant hotter objects. This then is often the first signpost to an area of recent star formation – the mere presence of high-mass stars. Modern techniques suggest that we include CO hot spots and radio continuum peaks in the same spirit, namely to trace the high-mass star-forming component of a region.

4.6.   Connections between stars and clouds

What other attributes should we anticipate in young stars? Stars form in dense pockets of dusty gas. One expects an intimate association between a young star and its parental cloud typified, for example, by the presence of bright nebulosity – glowing patches of gas, illuminated by the star, and obviously connecting star and cloud. Fig. 4.1 illustrates some examples of bright nebulosity associated with stars now known to be young. In Chapter 5 we shall treat, in some detail, the observed characteristics of young stars. For the moment we shall say simply that we would scarcely expect a very young object to resemble the appearance of a normal, mature star. There must be some finite period in which the fledgling star adjusts to its new circumstances; during that period we might well see spectroscopic peculiarities – perhaps an active exchange of gas between star and cloud, or some symptom of the initial collapse phase from cloud to star. The close association of young stars and dark clouds also suggests that much dust should attend the birth: perhaps thermal infrared emission from this dust should be sought? Our sun spins very slowly in the present epoch. Was it always so? Might not young objects have a radically different spin rate from that of mature stars? This, too, will be discussed in Chapter 5.

We have seen the basic ingredients that will be present in potential stellar nurseries: large amounts of gas, chiefly hydrogen, in both atomic and molecular forms; other molecules formed either in the dark cloud at large or on the surfaces of dust grains. Magnetic fields will also be present, threading the cloud in possibly intricate patterns, controlling the motion of any ionised gas (neutral atoms have no net charge for the magnetic field to push, but ions have positive charge since they have lost electrons, which themselves have negative charge). Rotation of the original cloud complex is likely to communicate itself to any piece of the cloud that attempts to collapse to form stars.

Thus far we have not addressed the issue of how efficient the process of star formation might be. There are two ways to pose this question. What is the minimum mass needed to produce a star of given final mass: is all the gas and dust utilised fully or is it necessary to supply much more material than will be locked into the finished stellar product? Alternatively, how much of the material in a dark cloud complex is to be converted into stars each

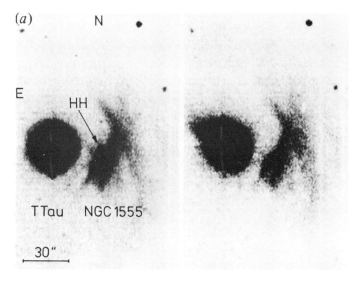

Figure **4.1 a)**   Two views of T Tauri (the big blob on the left of each panel) in different light. To its right lies the arc of Hind's nebula (here designated "NGC 1555"). The left panel is in the light of the sulphur atom, sensitive to "shocks" and it reveals a tiny gaseous blob ("HH") not seen in the right hand panel (taken in yellow light that is reflected starlight). (Photograph taken with the Calar Alto German-Spanish 2.2 metre telescope, courtesy of Thomas Buhrke and Reinhardt Mundt: Max Planck Institut fur Astronomie, Heidelberg.) For the significance of "HH" and "shocks" see Chapter 6.

time that these are generated? If star formation were 100% efficient, complexes would disappear once their first generations of stars were created. With only 10% efficiency a complex could make ten generations of stars before it would have exhausted its entire supply of interstellar material. At the low level of only 1% efficiency, dark clouds would survive for about one hundred

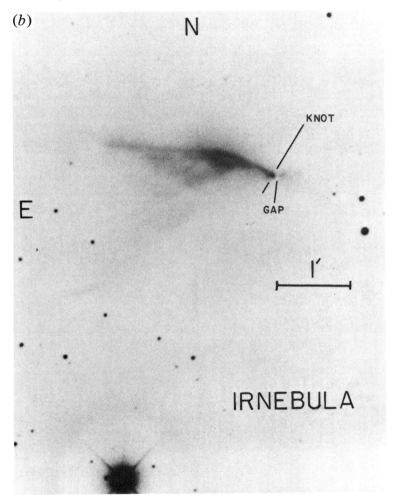

Figure **4.1 b)** The "Infrared Nebula" in the constellation of Chameleon. Its illuminating star lies very close to the knot. (Photograph courtesy of Cerro Tololo Inter-American Observatory, Chile and taken by Richard Schwartz.)

generations of young stars. In this context, a "generation" does not denote the entire lifecycle of any star but rather a period that typifies the ages of young stars that we can observe and recognise as such in a complex. Nevertheless it is clear that star-forming efficiency and longevity of dark clouds are inversely related.

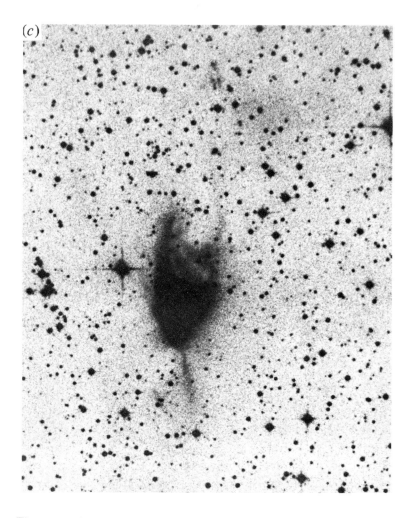

Figure **4.1 c)**   The fascinating variable nebula at R Monocerotis whose star lies close to the point at which the strange beam protrudes from the upper brighter fan nebula. (Photograph courtesy of David Malin, Anglo-Australian Observatory, with the permission of the Director.)

It is instructive to look at the overall structure of a stellar nursery since this will raise other questions about star formation. Fig. 3.8 shows a portion of the Taurus-Auriga dark cloud complex, one of the closest nurseries to the sun and one rich in T-associations. Even a cursory examination of the distribution of dark clouds (the relatively empty whitish voids on the photograph) suffices to show that the complex as a whole is not spherical. We see instead a series of blobs drawn out along a line, like beads on a wire. Each of these blobs represents the darkest, densest core of a more extensive molecular cloud which is visible only to carbon monoxide surveys at microwave wavelengths, not to light. Further, the blobs are interconnected by less dense, and much less opaque, dusty molecular material. What is the relationship, if any, between these dense cores? Does each signify a separate site for potential star formation? If one core formed stars, would its neighbouring cloudlets do so at the same time? What organises these cloudlets along an apparent axis? Perhaps the Galactic magnetic field?

Several of these cloud cores in Taurus-Auriga reveal obvious though faint stars, sometimes involved in wispy nebulosity. Certainly some recent star formation has taken place. But not all cores show signs of recent activity, as if each were independent of the circumstances of its neighbours. In Ophiuchus it has been found that the local magnetic field is definitely orientated parallel to this highly elongated nursery. Yet the overall pattern in Taurus suggests that the "strings" of cloudlet cores represent collapse of the original cloud into sheets, with the magnetic field perpendicular to the line of cores.

On the question of interactions between newly-formed stars and their ambient media we should consider two cases, one for O-, the other for T-associations. Take the example of the Rosette Nebula (Fig. 4.2). At the core of this rather regular ring of glowing gas filaments lies a cluster of blue O-stars. The combined radiation pressure (Chapter 3.2) of all these hot stars has acted to scour out the centre of the nebula, perhaps preventing the formation of any smaller stars by disrupting their prestellar dense gas fragments. For the lower mass stars we note that CO radio surveys are capable of providing the same velocity and intensity information as 21 cm hydrogen line studies. CO maps of the Taurus-Auriga complex reveal rapid motions in the gas in the immediate vicinities of a number of young stars (Chapter 6). These motions are presumed to be caused by stellar winds – flows of gas away from the surfaces of these stars that sweep out into the surrounding cloud and keep it stirred up. Do these vigorous spoons act to inhibit further small stars from forming in the clouds? We shall meet stellar winds again for they figure prominently in the litany of peculiar behaviours of young stars. Many massive, hot, very luminous stars drive such winds, probably by means of their radiation pressure. This particular mechanism is not available to appreciably less massive stars and there is much controversy as to a plausible means for low-mass stars to achieve their surprisingly powerful winds.

Figure **4.2**   The Rosette Nebula: a ring of gas blown out by the central cluster of O-stars in the nebular hole. (Photograph courtesy of David Malin, Anglo-Australian Observatory, with the permission of the Director.)

4.7. The "cosmic questionnaire"

We are now in a position to pose a series of simply asked, but not so simply answered, questions that raise several issues fundamental to star formation. This questionnaire comprises the following.

1. Can we find evidence of star formation spanning the entire range of masses that we believe can exist among stars?

2. What are the surface temperatures, radii, and ages of typical young stars once we become aware of them? These are clearly important questions for us since their answers must be used to help us understand our own origin and that of our solar system.

3. Do we understand the nature of the energy radiated by young stars at all wavelengths?

4. If we simplistically divide stars into two categories – "high-mass" (appreciably more massive than the sun) and "low-mass" (like the sun or even less massive), is it clear that these two types of object form in the same places; at the same time; even by the same mechanism(s)?

5. Young stars are observed to be obscured by dust and gas that intervene along our lines of sight to the stars. Exactly where does this obscuration arise (Fig. 4.3)? Is it strictly interstellar – between

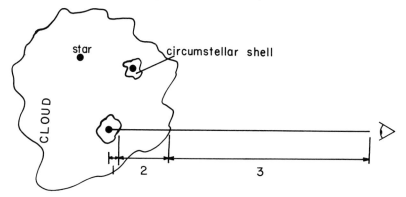

Figure 4.3 A dark cloud contains young stars, some with their own dust shells. The obscuration along the line of sight from the observer includes contributions from: 1) the circumstellar shell; 2) the intracluster medium (within the dark cloud between the embedded stars); and 3) the purely interstellar medium (between the surface of the cloud and the sun).

the sun and the surface of the dark cloud complex? Could it arise dominantly within the dark clouds? This we term "intracluster" since it represents obscuring material inside the clouds but between individual forming stars in the young cluster. Or, finally, might we be witnessing principally "circumstellar" envelopes – small dust clouds physically associated only with individual stars? Clearly one would expect some combination of all these three locations but perhaps one is more important than the others?

6. How efficient is star formation?

7. Do the stellar nurseries of which we are aware know with what frequencies to manufacture stars of different masses? In other words, do all nurseries produce the same mass spectrum or are low-mass and high-mass stars created in entirely separate regions of space?

8. Are there any observable peculiarities of young stars that distinguish them from more evolved objects? If so, how can we explain these peculiarities?

9. Does the birth of stars give us any hints as to how planets might form? In particular, can the study of young stars inform us on what timescale planets are made and in what kind of an environment?

10. How do already-formed young stars interact with their nurseries and do they have any effects upon the birth of future generations of stars within these clouds?

4.8. Detecting high-mass stars

It should be apparent from the above discussion that high-mass (hot, blue, luminous) stars are relatively easy to run to ground on the basis of radio continuum peaks, local peaks in CO heating of clouds, far-infrared enhancements, or plain old optical brilliance.

It is much more difficult to recognise their low-mass counterparts because there is such a gulf in luminosity between the two classes of star, hence perhaps in detectability by various techniques. The recognition of the low-mass stars will form the substance of Chapter 5 and if this seems chauvinistic, then it is! After all, our sun is a low-mass star.

5

How to recognise the youngest stars

all nearness pauses, while a star can grow
e. e. cummings

The only thing of which I could be certain was
that everyone had information to offer, spoke
in a voice that was full of authority, and
contradicted what the last fellow had said.
Norman Mailer,
"Ancient Evenings"

T Tauri stars – optical spectra – infrared – HR diagrams –
theory and observation confronted – stellar rotation –
ultraviolet and X-rays – stellar variability – magnetic fields

5.1. The T Tauri stars

We have already described the importance of initial stellar mass to the rapidity of subsequent evolution. It is this that renders the hot, blue, high-mass stars valuable indicators of recent star-forming activity since they have had little opportunity to drift far from their birthplaces. If we are alerted to the likelihood that a particular region of dense, dusty clouds contains recently-born high-mass stars and, therefore, perhaps young low-mass objects too, how are we to recognise these latter fledgling suns?

During the 1940s, Alfred Joy, in California, made a number of surveys of dark clouds, seeking stars that showed a bright emission line of hydrogen gas. Now the mere presence of such a line does not automatically imply the youth of a star. Indeed, it is a characteristic of a very broad range of celestial objects, among which are the likely phases through which our sun will go, long after it passes middle age and leaves the main sequence. However, by looking specifically in dark clouds, Joy greatly restricted the probability of encountering these post-main-sequence objects and enhanced the prospects of finding pre-main-sequence stars. He isolated a distinct class of stars satisfying several basic criteria, all of which suggest that these are members of the Galactic population of low-mass young stars.

The prototype of these stars first made its debut in the annals of astronomy in the mid-nineteenth century when the astronomer Hind noted it as an object intimately involved with bright, glowing, nearby gas that seemed to curve in a graceful arc around the star (Fig. 4.1a). It was the star T Tauri, in the constellation of Taurus the Bull. The designation of a star by a letter in this manner indicates that the light output of the object has been observed to vary with time. The brightness of the star itself, and of Hind's Nebula which reflects the light of the star, were highly variable during the nineteenth and early twentieth centuries. In 1890, Burnham, observing T Tau, found another bright nebulosity very close to the star. The new nebula underwent a major outburst in 1897, that was observed by Barnard. It is particularly interesting to note that the history of T Tau clearly demonstrates the value of archival documentation. A comparison between very old images of the star and its attendant nebulæ, and very recent ones, shows that the tiny blob of material identified as "HH" in Fig. 4.1a (much more about "HHs" in Chapter 6) has actually moved away from the star in the course of the past 70 years. T Tau may in fact be a poor prototype of its class of stars because it displays so many strange behaviours but we shall discuss these later.

The primary criteria of Joy's class of stars were: 1) association with bright (glowing gas) or dark (obscuring dust) nebulosity; 2) erratically variable light output; 3) bright emission lines of hydrogen and calcium, among others; 4) when recognisable, a stellar surface temperature cooler than that of the sun. Another distinguishing characteristic of these stars is the high abundance of lithium in their atmospheres. Now lithium is rather fragile – it is easily eaten up inside stars by nuclear reactions. Therefore its presence is an independent clue to the extreme youth of the T Tauri stars. Theory leads us to expect that stars come to maturity on the main sequence from the region above and to the right of it in the Hertzsprung-Russell diagram. Might these T Tauri stars represent the precursors of stars like the sun?

Joy also noticed that the features of the stellar absorption spectra representing cool, outer layers of the atmospheres seemed rather washed out compared with normal (mature) stars of the same temperature. Were their features all to be broadened, relative to normal stars, this would account for the difference. Broadening of spectral features can arise from several different physical causes, one of which is rapid stellar rotation (Chapter 5.2). This, as we shall see in Chapter 5.6, is definitely an expected diagnostic of a young star.

5.2. Optical spectra

Quantum mechanics (Chapter 3) tells us precisely the frequencies at which different atoms, ions or molecules will make their transitions. We should distinguish clearly between two types of situation. An atom, sitting in cold space, bathed in the radiation field of a warm star, will eagerly grab those photons that enable any of its electrons to be excited to higher energy levels. The atom is said to "absorb" radiation at these frequencies. Conversely, a hot excited atom, sitting above a cool stellar atmosphere, will radiate away the energy that its electrons have, above the ground state, and will add its radiation to that coming from the stellar background. This atom is said to "emit" photons at these frequencies. When we look at the spectrum of starlight, dispersed through a prism, for example, from our scrutiny we can sometimes see two kinds of spectral feature. If the spectrum is a generally bright continuum of light, there will be various dark bands (regions of less than average intensity if we plot energy received at different wavelengths) that cross the continuum. These are absorption features, the pattern of whose frequencies will enable us to identify the responsible atomic element, ion or molecule. Further, the atoms that appear in absorption, and the depths of their absorption lines (darkness of the bands) are uniquely characteristic of a single temperature for the star's outer layers – its "photosphere" (Fig. 5.1). Similarly we may note regions of enhanced brightness in the spectrum. These are emission lines, corresponding again to specific quantum mechanically predicted frequencies.

Stars are hot at their cores (even pre-main-sequence objects and some phases of protostars) and usually much cooler in their outer atmospheres. Therefore we can identify the absorption lines in a stellar spectrum with atoms in the outer layers of the same star, for the radiation from inner, hotter atmospheric regions must run the gauntlet of passage through the more distant gas layers to reach us. Similarly, if we see, superimposed on all this background of continuous radiation and absorption features, a number of bright emission lines then we must conclude that, for some reason to be explained, there is anomalously hot gas above the star that contributes its own emission to the overall stellar spectrum (Fig. 5.1).

More than this, we can examine "profiles" of the lines (bright or dark), that is, their pattern of brightness (i.e. energy radiated) with wavelength (and wavelength is related to line-of-sight velocity of the gas atoms via the Doppler effect: Chapter 3.2). If the absorption lines are very sharp and narrow we know that this

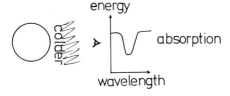

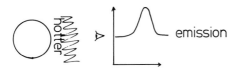

Figure 5.1   A star (left-hand circle) has gas above its surface. Top panel: this gas is colder than the stellar surface layers so it absorbs stellar radiation, leaving an absorption line. Lower panel: hotter atoms do not absorb stellar radiation but contribute their own emission line to the overall stellar spectrum.

implies a star whose atmosphere is very extended (a "giant" star – the stellar analogue of an obese person – a star with seemingly too much luminosity for its temperature); deep, stubby line profiles connote a "dwarf" – a much smaller star with a correspondingly more compact atmosphere. If all the lines, bright or dark, are shifted in wavelength from the laboratory wavelengths, we can determine the radial (along our line of sight) velocity of the star. Lines that are broadened, in an appropriate and quite specific fashion, can indicate that our star is in rapid rotation, thereby blurring the normal sharpness of the absorption features. If emission lines have a great width then we can infer that the responsible emitting atoms occur in a region with a wide range of velocities, to cause the Doppler breadth that we see. Finally, and rather more exotically, if the lines are seen multiply (if each basic line is flanked by weaker neighbours with the correct spacings in wavelength) this can inform us of the presence and even the strength of sizeable magnetic fields on the stellar surface (not those of any nearby clouds or interstellar medium, however, unless we are observing lines from the clouds themselves in the radio region of the spectrum). These are the basic ingredients of stellar spectroscopy and their interpretations.

Let us examine the T Tauri stars in more detail. First, their optical spectra can be surprisingly complex. Consider Fig. 5.2 which shows a montage of stellar spectra all selected to represent objects with 4000 K photospheres. At the bottom we see a normal, mature star "61 Cyg B" which reveals an abundance of absorption features. These plotted spectra represent the intensity of radiation at different wavelengths so that emission features show as pronounced high peaks in the tracings, and absorptions appear as troughs in the general background level. For example, note the prominent sharp trough near 5895 Å which is absorption due to cool neutral sodium atoms in the outer stellar atmosphere (this is the same orange line that often illuminates our main roads at night when it is used in emission lamps). Moving up the page, from 61 Cyg B, notice that there are fewer and fewer troughs (absorptions) and correspondingly more peaks (emissions) until RW Aur seems to show all the absorption lines of 61 Cyg B but in emission. These are all spectra of T Tauri stars, except for that of 61 Cyg B. Clearly the phenomenon responsible for making the spectra so complex can manufacture either mild or extreme versions of T Tauri stars. All, however, are recognisable from their spectral peculiarities. Notice also the line near the right edge of these spectra, at 6563 Å, which is so powerful that it has been truncated so as not to obliterate details in the spectra above on the page. This is the "H-alpha" line of hydrogen in emission that Joy utilised to seek his strange objects in dark nebulæ: its prominence is obvious.

The richness of the spectrum of the extreme T Tauri star, RW Aur, is evident. In addition to those of hydrogen, lines of helium, iron, magnesium and sodium all can be recognised. Joy observed that these rich spectra were reminiscent of the bright-line spectra of the sun's upper atmosphere which became visible during total solar eclipses. This region of the solar atmosphere, above the photosphere (the surface layers), is termed the "chromosphere" because of these bright emission lines. The chromosphere represents an outer region of the sun in which the temperature increases outwards, rather than falling off as occurs within the

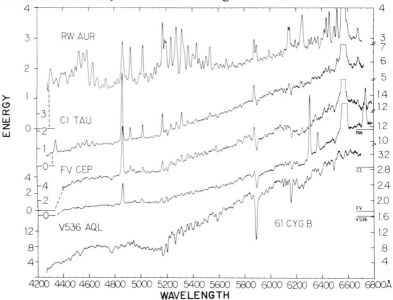

Figure 5.2   A montage of 4000 K stellar spectra. 61 Cyg B is a normal, mature star; above it are the spectra of 4 T Tauri stars showing the emission-line spectrum common in these stars, with increasing strength up the page.

volume surrounded by the photospheric layer. We believe that the same physics describes T Tauri stellar atmospheres, although this analogy does not explain to us what chromospheres have to do with potential pre-main-sequence objects.

5.3. Infrared observations

It is their intimate relationship with the parental dark clouds that provides the strongest morphological evidence that T Tauri stars are recently born. Not all the material of a contracting fragment of cloud will find its way into a protostellar core. Much debris will attend the birth and it is precisely the gas or dust of this debris that we can recognise in the form of nebulosities that bind new-born stars to their nurseries (Fig. 4.1a,b,c).

A consequence of this still-close connection between cloud and star would be the presence of substantial dusty material close to the star; not at the low temperatures characteristic of dark clouds without internal heating sources in the form of embedded young stars, but at higher levels that argue for appreciable thermal input of a nearby star. In this respect, too, T Tauri stars do not disappoint us. We find not merely that it is useful to observe these objects at infrared wavelengths, which are sensitive to warm or hot dust (a few hundred up to just over 1000 K), but even that some T Tauri stars seem to radiate more and more energy as we lengthen the wavelength of our scrutiny. Fig. 5.3 shows the energy distributions of a sample of young stars, illustrating the steeply rising character of some stellar spectra (notably that of HL Tau, for example: we shall meet this star again later!).

For a short period of time (about 3 years), there was some controversy as to the origin of all the extra infrared emission that could not be legitimately attributed to the stellar photospheres. You may wonder how anyone could have viewed these steeply rising spectral energy distributions and seen any resemblance to stellar photospheres in them. While it is true that typical stellar surface temperatures rank in the thousands of degrees and should, therefore, produce energy curves that decline with increasing wavelength, we are perhaps neglecting a crucial factor. The role of dust in extinguishing starlight should be thought through with care.

Let us consider the effect of passing starlight through a dust cloud without worrying about the precise location of this cloud. We know that dust attenuates the intensity of starlight but exactly what does this mean? Imagine a stellar photon – a packet of starlight energy – entering a cloud of dust. The photon is characterised by its energy; this in turn, gives it a characteristic wavelength or colour. For example, an X-ray photon is highly energetic and extremely short in wavelength, whereas an infrared photon is not very energetic and long in wavelength. By and large, interstellar dust grains prefer to absorb the high energy of shorter wavelength photons. The grains are often comparable in size to, or even larger than, the wavelengths of these photons. For such a photon, there is

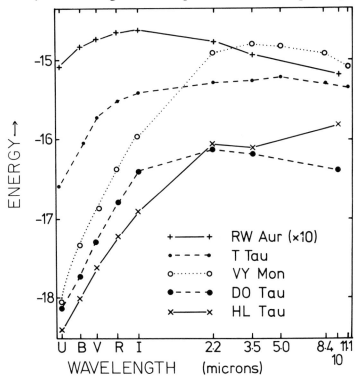

Figure 5.3   The importance of making infrared observations of T Tauri stars is illustrated by the energy distributions of five such stars, most of which show increasing luminosities at long wavelengths.

no escaping the grain, whose intention it is to absorb totally the energy of the photon at its surface, thereby contributing to the thermal energy of the dust grain. Short wavelength starlight literally heats grains because grains are efficient in absorbing them. By contrast, long wavelength weak photons have large wavelengths compared with typical grain dimensions. Such photons scarcely break stride as they encounter these relatively tiny dust particles. Dust is much less efficient in capturing infrared photons than optical (or, especially, ultraviolet) photons. Larger wavelength radiation passes through a dust cloud largely unscathed. There is a second process, distinct from pure absorption, that nevertheless greatly enhances the difference between the probabilities that short and long wavelength photons are captured by dust particles. This process is termed "scattering". Imagine yourself to be an elephant, sauntering along a beach littered with inflated beach balls. Your great size, compared with the diameter of a beach ball, renders your route across the beach one that is relatively independent of the presence and frequency of beach balls. However, a mouse, intent upon the same beach stroll would find his passage considerably interrupted and modified by the existence of these balls. His actual route would show profound deviations from his intended path: he would be "scattered" by the balls. In like manner, blue photons are scattered from their otherwise straight line courses by encounters with sufficiently large particles or molecules, whereas infrared photons travel much more directly through dusty regions. A direct consequence of scattering is that every short wavelength photon must run the gauntlet of true absorption by grains many times, as it zigzags its way, back and forth, trying to exit from the cloud (c.f. Chapter 2.1).

In short, if we try to pass a spectrum, in which all frequencies are equally represented, through a dust cloud, what emerges of the original beam is principally the longer wavelength component (Fig. 5.4a). In fact, if we view even a very hot star, rich in ultraviolet radiation, through a sufficiently dense cloud, only red light emerges (Fig. 5.4b). This "reddening" effect we call the "extinction" of starlight, and it consists of the two independent, but augmenting, processes of scattering and absorption. But what of all this absorbed, short wavelength energy that is tapped by dust grains? This energy heats grains to temperatures above their ambient levels in the interstellar medium, and causes them to re-emit the absorbed energy thermally in accordance with these temperatures. Now we need to distinguish two common situations (Fig. 4.3), in which the dust cloud that intervenes between a star and ourselves is either a truly interstellar cloud or a circumstellar cloud, physically associated with the star. In the former situation, the intensity of starlight is greatly diminished (in accordance with the inverse square law) by travelling far through space. Consequently, interstellar grains gain extremely little energy by their preferential absorption of high energy starlight and their reradiation takes place at very far-infrared wavelengths. However, in the case of a circumstellar dust cloud, the grains are capable of absorbing an appreciable fraction of the total starlight, depending upon their physical properties and their total numbers. This starlight raises the circumstellar grains to temperatures of hundreds, even a thousand or more degrees, producing reradiated thermal emission in the near-infrared part of the spectrum. It is this augmentation (Fig 5.4c) of an extinguished stellar spectrum by reradiated infrared emission that identifies the location of the dust as circumstellar.

Consequently, it was not foolish to wonder whether the steeply rising spectra of T Tauri stars derived from circumstellar dust or just from the passage of starlight through interstellar clouds. Actually the real controversy went even deeper than this. Some astronomers felt that a purely gaseous emission process could

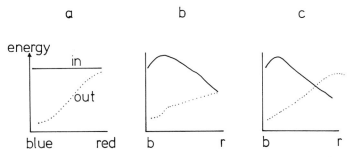

Figure 5.4   a) The emergent (dotted: "out") energy distribution of a flat input spectrum (solid: "in") to an interstellar dark cloud. b) The same but for an energy distribution more characteristic of a real star. c) The emergent energy spectrum from a circumstellar dust cloud.

contribute significantly to the extra infrared radiation of T Tauri stars. We have already encountered the process of ionisation (Chapter 3.1), by which an atom is stripped of one or more of its originally bound electrons, principally due to the capture of sufficiently energetic photons which can provide the necessary energy to unbind an electron, or electrons. Even such a freely travelling electron is not immune from encounters with ions other than the atom from which it was originally liberated. Without ceding any freedom, free electrons can be influenced by these positively charged ions and curve around them (Fig. 5.5), losing energy in the encounter, which is radiated away. By virtue of the electron's free status (unbound to any atom or ion before and after these encounters) we term the emergent radiation "free-free emission". It was this process that some people felt could explain the near-infrared brightness of T Tauri stars.

There is, however, an acid test capable of distinguishing between this type of gaseous emission and dusty thermal emission. It involves a tool new to us but very useful in astronomy, called a "two-colour diagram". Consider yourself a primitive astronomer, obliged to view the Universe through any one of three coloured filters designated U (ultraviolet), B (blue) and V (visual). How could you recognise the existence of stars of very different surface temperature, and the main sequence, using only these three filters? You could draw the two-colour diagram using the differences in amounts of energy radiated at each of a pair of wavelengths, for

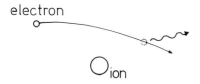

Figure 5.5 "Free-free" radiation from a free electron curving around an ion.

example U and B, or B and V. Fig 5.6 illustrates the (U-minus-B, B-minus-V) plot for stars on the main sequence.

We have labelled the sense of each axis by the simple terms "blue" and "red", meaning more radiation through the U filter than B (or B than V) and less radiation through the same pairs, respectively. We would find that most stars lie on a curved locus through our diagram that must represent the main sequence. Stars in the top left are the hottest; in the lower right are the coolest. We could equally have constructed our diagram using three filters located at different wavelengths in the infrared part of the spectrum. Our infrared two-colour diagram would also include the main sequence and would show "hot" and "cool" infrared emission in different locations. The merit of our infrared diagram is that it is very sensitive to emission other than the simple thermal radiation of stars. Suppose we have a star which heats some dust particles in its circumstellar environment, by using its own starlight. This star would now be in a location off the main sequence in the infrared two-colour diagram ("A" in Fig. 5.7). Add more circumstellar grains (all at the same temperature) and the star travels out from its original main sequence location along a trajectory (A-A) in our plot, that terminates when we no longer

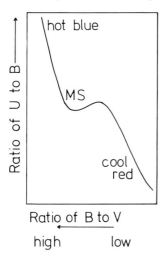

Figure 5.6   Optical two-colour diagram showing the trend, between ultraviolet-blue and blue-visual colours, that represents the main sequence ("ms").

recognise the object as a warm star but as an entity radiating all its energy as a cool blackbody. Do the same to another star, but use grains that lie in a cloud at a much greater distance from the stellar surface so that they achieve a rather lower grain temperature. Now the star moves along trajectory B. So, we have only to observe a star in the three infrared filters, compare its intensities pairwise, and locate it accordingly in Fig 5.7, to understand whether it is an ordinary main sequence star, or a star possessing its own dusty envelope, in which case we can even hazard a guess as to the temperature of a typical grain in that envelope. What of free-free

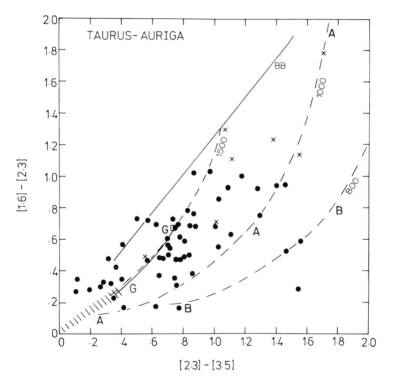

Figure 5.7   Infrared two-colour plot for T Tauri stars. Hatched (lower left) region is where normal stars lie. Adding the thermal emission by hot dust grains (temperatures 1500, 1000 and 800 K are illustrated) to the stellar photospheric radiation moves stars along the dashed tracks to the upper right. Free-free (gas emission) moves stars along the solid curved track (G) but only as far as the open square. A number in square brackets (see axis labels) denotes a measure of the intensity of radiation at a wavelength of that number of micrometres.

emission? That would move a star outward along another characteristic track in Fig. 5.7 (G-G), terminating as indicated. What does all this have to do with T Tauri stars, you may ask? Fig 5.7 presents a real two-colour diagram for about 60 T Tauri stars in the relatively nearby Taurus-Auriga nursery. See how many stars are located far from the location of normal stars (bottom left corner) and far away from the track of gaseous (free-free) emission. Only the presence of thermal emission from circumstellar dust grains could explain the positions of most of these stars in this diagram. In fact, there was never any dispute as to the dusty origins of radiation in the middle- and far-infrared regions of T Tauri spectra. Only that interesting realm of the near-infrared (wavelengths between 2 and 10 times as long as yellow light), where the T Tauri stars begin to rise above the expected radiation from normal stars of their surface temperature, caused the controversy. Gas emission may well occur in some, even all, of these stars but the dominant source of this once unexpected near-infrared emission is not due to any free-free process. Only the existence of appreciable amounts of circumstellar dust can account for these energy distributions.

5.4.   A Hertzsprung-Russell diagram for T Tauri stars

In Chapter 2.4 we mentioned the necessity for choosing an appropriate set of theoretical calculations with which to interpret the observations of young stars, and the important role that the HR diagram would play in this choice. What is required to construct such a diagram for the T Tauri stars? We need to focus our attention on a single complex of clouds of gas and dust in which T Tauri stars occur, preferably one whose distance is known. The number of T Tauri objects should be substantial and the nursery relatively close to us so that we are not hampered by having selected only the brightest of these stars in the complex. The Taurus-Auriga dark cloud complex, in which Joy conducted his original hydrogen-alpha surveys, provides an ideal testing ground. Lying only 500 light years away (compared, for example, with almost 1,500 light years for the rich star-forming regions near Orion's Great Nebula), it contains a hundred T Tauri stars and is therefore well-sampled by various techniques. We basically need to know stellar surface temperature and total luminosity. The former is accessible solely through optical spectroscopy (see Chapter 5.2), the latter through a combination of optical and infrared observations. Of course, we should understand that a few stars, like RW Aur in Fig. 5.2, may yield no useful temperature information to optical spectroscopy due to the extremely rich emission-line spectrum. However, we lose only 8 stars in the Taurus-Auriga clouds to this problem. The culmination of optical and infrared studies for Taurus-Auriga is symbolised by Fig. 5.8 which is the HR diagram for this stellar nursery.

5.5. Comparison with theory

The disposition of stars within the diagram suggests that only convective-radiative tracks (Chapter 2.4; Fig. 2.5) provide a satisfactory interpretative framework for the observations (since dynamical models do not suggest young stars to be optically visible in those parts of the diagram most densely occupied by the observed T Tauri stars). Given that this is the case, a diagram such as Fig. 5.8 may be interpreted to yield quantitative information on basic stellar parameters (radius, mass, age) by using an overlay like Fig. 5.9. This figure depicts the evolutionary tracks for stars with core masses from 0.1 to 3 solar masses (the heavy lines). A 1.25 solar mass core will evolve along the track labelled "1.25" from right (high on its convective track) to bottom left (the radiative track which terminates with the onset of hydrogen burning in the stellar core, shortly before the arrival of the star on the main sequence. As this star evolves, its track cuts the several lines of constant stellar radius (dashed) so that, earliest, it has perhaps 10 solar radii; at a later stage only 3 solar radii; and it finally reaches the main sequence with close to the sun's present radius. This contraction is in accord with expectations so there is nothing mysterious couched within the complex detail of Fig. 5.9. Finally, what do the curved dashed lines in this figure represent?

Suppose we began the evolution of a cluster of stars, all with different masses, at a common instant. How would successive snapshots of the cluster appear? The more massive stars would always move more rapidly than their lighter contemporaries and, hence, would be seen closer to the end of their tracks. Therefore we can define a series of "isochrones" (from the Greek, meaning equal times) that represent the locations along appropriate evolutionary tracks of stars of different mass, viewed at common times into their evolution. The shapes of these isochrones would be as indicated by the dashed trajectories in Fig. 5.9 which are labelled from A

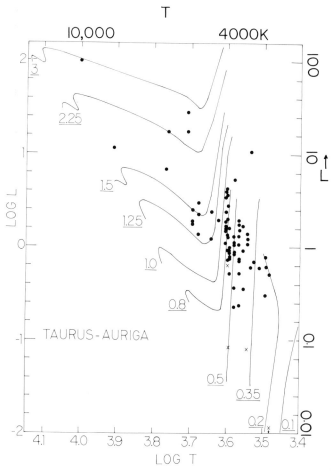

Figure 5.8 The HR diagram from the Taurus-Auriga nursery, showing convective-radiative tracks labelled by the mass of the stellar core (in solar units). Right-hand axis is in units of solar luminosity; left-hand is logarithmic, also in solar units. Top axis labels temperature linearly; bottom axis is temperature, logarithmically.

(earliest time) to H (latest time). By interpolation between the grid lines of evolution, radius and isochrone we can discover these basic stellar characteristics for any star that we can locate in the HR diagram on the basis of our optical and infrared observations.

Incidentally, Fig. 5.9 does illustrate one major difference between the evolution of stars of the lowest mass and those of more moderate (around 2 solar masses) mass. Very low-mass stars evolve essentially vertically in the HR diagrams; that is, their evolution consists entirely of the convective phase, and even this requires very substantial periods of time before the main sequence is attained. (This explains why "failed stars" or brown dwarves, like Jupiter, spend 4,500 million years evolving – since the solar system was formed – yet still never have attained the main sequence.) By

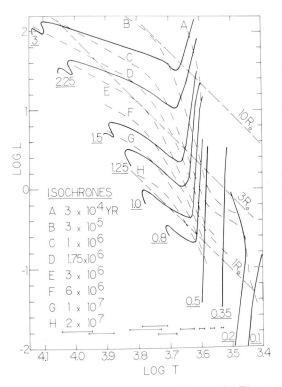

Figure 5.9 Interpretative HR diagram for Fig. 5.8 showing evolutionary tracks, lines of constant radius (dashed lines), and isochrones (curved dashed lines denoted by letters). R-subscript-sun-symbol denotes radius in solar units.

contrast, stars appreciably more massive than the sun undergo relatively brief convective phases and spend the greater proportion of their pre-main-sequence lives on their radiative tracks. This difference will prove useful later when we try to interpret observations of stellar rotation in detail.

Recently, there has been a very pleasing marriage of theory and observation that can be interpreted as vindication of both the observations that went into the construction of the HR diagrams of stellar nurseries and the basic theoretical framework within which

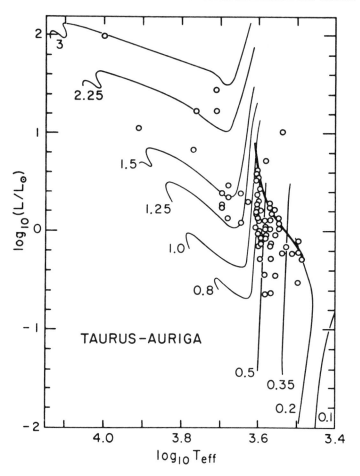

Figure **5.10**   HR diagram for Taurus-Auriga showing the calculated birthline (solid heavy curve cutting the convective tracks at centre-right).

the observations have been interpreted. This remarkable connection comes about when one asks a simple, but important, question of the T Tauri stars. At what location in the HR diagram do stars with different masses first become visible? This relates basically to the amount of material still shrouding the star as it evolves, and the question addresses the issue of when it is first possible to see through the remnant cocoon of dust and gas and to recognise a stellar surface or atmosphere. The term used to describe the locus on the HR diagram of stars of different mass as they achieve this moment of initial visibility is the "birthline". Now how do the observations stack up against the purely theoretical birthline? Fig. 5.10 illustrates the HR diagram showing our 100 T Tauri stars in the Taurus-Auriga nursery, together with the calculated birthline. There is a remarkable upper envelope to the distribution of stars actually observed that mimics closely the birthline! This accord can be taken to imply that the necessary calculations and assumptions that have been employed to transform the original visual and infrared observations into a location for each star in the diagram are in some sense "correct". It also corroborates the basic set of calculations from which the birthline comes, which is a set of spherical collapse models for protostars, that suggests the essential relevance of the convective-radiative approach. This is the kind of "discovery" that is capable of persuading even some sceptics that we are on the road to understanding T Tauri stars. Of course, there are always sceptics, even now...

5.6. Stellar rotation

As we shall see in Chapter 10, it is critical to know the history of a star's rotation rate in order to investigate whether or not it has produced a potential planetary system. How can the spin of a star be measured? Now "spin" really means "angular momentum". What is "angular momentum"? Any object that rotates has angular momentum and this must be conserved. You all know those thrilling spins that ice skaters include in their repertoire – the spins in which the skater becomes a mere blur of limbs and colours. How do they achieve this great speed? Watch them carefully next time and notice that during slow spins their arms or legs are outstretched but when they hug their limbs to their bodies then they rotate much more rapidly. That is the conservation of angular momentum. (It is the combination of spin speed and compactness, or otherwise, of the mass distribution that dictates the angular momentum: the product of mass, velocity of the mass, and its distance from the axis about which everything is spinning.) Stretch out your arms and some of your mass is now far from the axis of rotation – you rotate slowly. Pull them in and now all of your body is close to the spin axis and you must rotate rapidly to maintain the constant value of momentum that you had before. Essentially, then, by watching the trend of stellar angular momentum with time we can infer something about the compactness or extensiveness of material both in the star and that has been shed by the star.

Our basic information on stellar spin can come from one of two methods. First, let us suppose that the surface of our star is spotted as is the sun itself. There may be areas of the surface that appear much darker than the rest of the disk because these areas are a little bit cooler than the average photospheric temperature. Sunspots are not really black; they simply appear that way by contrast with the bright disk. If you are wondering how much cooler such a spot should be then remember (Chapter 1.6) that the total luminosity radiated by any hot object is proportional to the fourth power of temperature (i.e. temperature multiplied by itself three times). Take the sun, with an average photospheric temperature close to 6000 K. How much cooler would a spot be, that radiated only half as much as a neighbouring patch of the solar disk of the same size? The answer is, about 1000 K cooler, which is not a great deal. Likewise, were our star to have exceptionally bright regions, where the temperature was somewhat elevated above the average for the surroundings, these bright spots would show up as local enhancements of brightness. Let's measure the total brightness of this star on a regular basis – perhaps every couple of hours – for several nights. We might find a rather regular variation of total starlight (Fig. 5.11). This, we might recognise, was a symptom of a spotted star that was rotating every few days, carrrying the bright (and/or dark) spots across the stellar disk. Although we certainly cannot spatially resolve these minuscule spots on an already minuscule stellar disk (except for the sun and an occasional "supergiant", or enormous star, like Betelgeuse in Orion), nevertheless their local effects on total brightness of the star can be recognised. Knowing the period of the stellar rotation, we can readily calculate the actual surface rotation speed since the size of the star is known from its place in the HR diagram.

The second method for measuring stellar rotation is based on direct observations of the stellar spectrum. In Chapter 5.2 we described the nature of emission and absorption lines as due to the escape or capture, respectively, of photons by atoms at precisely determined frequencies. As you can tell from Fig. 5.2, real stellar spectra don't seem to have infinitely narrow lines, either in emission or absorption. Now there are two basic reasons for this. Spectrographs always represent a compromise in design, which can lead to a compromise in derivable science. You can have an instrument that will subdivide starlight into very tiny pieces, yielding "very high resolution spectra". Of course, this presupposes that you can get enough light from your star to split up so finely and still be able to detect the photons. Or, you might prefer to study much fainter stars but not demand so high a spectral resolution. The spectra represented in Fig. 5.2 came from a highly sensitive but relatively modest resolution spectrograph. That was just what was needed to make an adequate HR diagram: enough spectral resolution to be able to recognise temperature-sensitive absorption lines but enough sensitivity to study a large

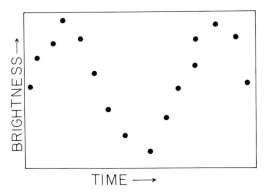

Figure **5.11**   Variation of brightness of a spotted star.

sample of stars, bright and faint, in a single stellar nursery. So the width of a line in a stellar spectrum is determined by several very different considerations: the intrinsic stellar parameters (rotation speed, extent of the atmosphere, presence of prodigious magnetic fields); the resolution of the spectrograph used; even the atmospheric conditions at the time the spectrum was obtained. Given adequately fine instrumental resolution it is possible to recognise real differences between the shapes (or "profiles") of stellar lines that are due to the size of the star's atmosphere, its temperature structure (how the temperature varies as you climb away from the surface of the star), its rotation rate, even the presence and strength of magnetic fields, and of stellar winds

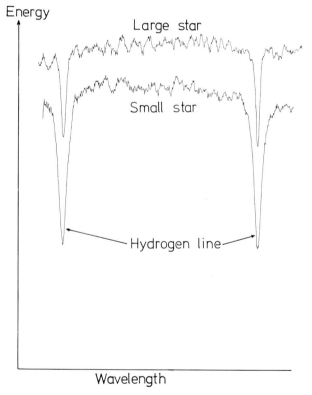

Figure **5.12**   Hydrogen absorption lines in two stars with identical surface temperatures; the upper is a very large (supergiant) star; the lower a main sequence (small) star. Note the marked differences in line depths and widths.

(concerted outflows from the surface like the solar wind but usually much more powerful). To illustrate the way in which line profiles can reflect these physical differences let's look at Fig. 5.12, which shows the profiles of two hydrogen absorption lines, both referring to stars of the same photospheric temperature (about 9600 K), but one small and the other large in radius. See how much narrower and sharper is the line from the supergiant (big) star. Likewise, high rotation speed betrays itself by a change in the line profiles and Fig. 5.13 presents the identical lines in absorption, one from a slowly rotating star, the other from a much faster rotator. You see that high rotation is identifiable from the broadening of the line profile. To extract the value of the rotation speed is not an easy matter; you must choose with care your star, spectrograph, wavelength of line, and method for determining the line width and shape. But the technique does work and it has been applied with some success to the T Tauri stars.

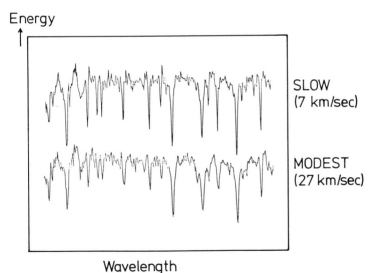

Figure **5.13**   Absorption lines in stars of the same surface temperature but with different rotational speeds. The technique is powerful enough to distinguish even relatively small differences. See how much broader are the absorption lines in the spectrum of a star with even a modest velocity compared with a slow rotator. (Spectra courtesy of Claude Bertout, Institut d'Astrophysique, Paris.)

The results, however, have been somewhat surprising. Fig. 5.14 presents the HR diagram for T Tauri stars in our favourite nursery – the Taurus-Auriga dark cloud complex – where stars are symbolised in accordance with their rotational speeds (large symbols for fast rotators; small for slow ones). Simplistically we can predict what we expect to find. Figs. 5.8 and 5.9 tell us that stars contract as they evolve down their convective tracks. Since they began life as greatly extended clouds this does not surprise us at all. But, as a star contracts, it must "spin up" – in other words, its rotation rate should increase to preserve angular momentum. So the older, smaller stars should be spinning faster than the younger,

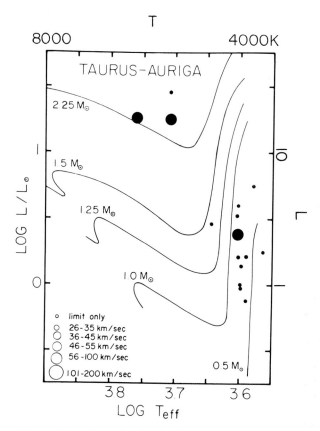

Figure **5.14**   Rotational velocities in the HR diagram for Taurus-Auriga. Sizes of circles represent the magnitude of the rotational velocities of stars (large symbols for fast rotators; small for slow ones).

larger ones. What we actually find is that the older and smaller T Tauri stars are rotating too slowly to show appreciable broadening of their spectral lines! Most of the more massive objects studied do rotate rapidly but that doesn't explain the curious behaviour of the T Tauri stars; nor does the observation that low mass stars aren't spinning that much more rapidly at the tops of their convective tracks than they do significantly lower down these tracks! It seems that the only way out of this problem is to admit that, somehow, the T Tauri stars have already solved their personal "angular momentum problem" by the time that they have become visible to us. We shall return to this apparent mystery, with a potential and exciting explanation, in Chapter 10.

5.7. Ultraviolet and X-ray observations of T Tauri stars

During a total eclipse of the sun, the disk of the moon covers the photosphere of the sun, yet much of interest and spectacular appearance is seen above this surface layer of the solar atmosphere. Immediately above it lies the chromosphere – a thin zone that produces bright emission lines, rather than the absorption features seen in the photosphere. (It is interesting to note that the element "helium", a name derived from the Greek word "helios" for "sun", was first identified in the solar chromosphere from its emission lines.) We know from Chapter 5.2 that bright lines indicate the presence of atoms in the chromosphere at higher temperatures than those in the photosphere. So the solar temperature must be increasing as we move upwards from the photosphere into and through the chromosphere. Beyond even the chromosphere lies the tenuous, feathery, subtle solar corona with its roughly two million degree temperature. (It is the solar corona, with its outreaching, roughly radial, filaments that renders the sun so spectacular in appearance during a total eclipse.) Remember that, because temperature is equivalent to velocity of particles, the corona consists of very fast-moving gas. It is so rarefied that, whatever the energy source for heating the corona, a little energy distributed among so few gas ions goes a long way; hence these great speeds and temperatures. Indeed, so fast are these elements moving that they escape entirely from the sun's gravitational influence to produce what we know as "the solar wind".

Now where in the spectrum would gas between the photospheric 6000 K and the coronal 2,000,000 K emit most copiously? Clearly, it is crucial to study the ultraviolet and X-ray emission from stars to seek evidence for the existence of chromospheres and coronæ.

The T Tauri stars have yielded a great many interesting results to remote scrutiny from the IUE ("International Ultraviolet Explorer") – a satellite, above the earth's atmosphere, with modest ultraviolet spectroscopic capabilities. Indeed T Tauri stars must have very "souped-up" versions of the solar chromosphere judging by the relative brilliance of their ultraviolet lines that can attain 1000 times the luminosity of the same lines in the sun. Even typical T Tauri stars with only 4000 K photospheres show lines that come from 200,000 K zones. It is hard to avoid the conclusion that the surfaces of T Tauri stars are covered by an extensive network of substantially-sized active and quiet zones (bright and dark spots) that provide a vigorous and complex chromosphere.

Many T Tauri stars have also been observed in X-radiation, again by way of a space observatory; most are detected. Here, too, their properties are enhanced over the X-rays from normal main sequence stars of the same temperature, by perhaps factors of order 1000. If this energetic radiation comes from a corona, then perhaps its temperature is not as high as in the solar corona. T Tauri stars seem to sustain 600,000 K coronæ rather than ones in excess of a million degrees, but the actual temperature may be less important than the total energy involved in maintaining the coronæ. One odd finding relates to that most interesting subset of T Tauri stars with strong optical emission-line spectra. Whatever extreme conditions create these rich optical spectra are insufficient to guarantee X-ray detectability. It is interesting to speculate on why these apparently very active stars fail in this aspect. But first, in spite of the obviously energetic character of X-ray photons, it is important to recognise their tremendous vulnerability to capture and absorption by atoms that lie between us and the sources of X-rays. Every electron bound to an atom is delighted to encounter one of these energy-rich photons, leading to photo-ionisation. Suppose that T Tauri coronæ drive stellar winds, akin to the sun's but more vigorous. All that surrounding gas lurks around in the circumstellar environment of these stars. Consequently, no stellar X-rays may emerge from the immediate gaseous envelope of these stars: all may suffer local absorption. Alternatively, T Tauri stars may simply be incapable of sustaining both a corona and a strong chromosphere, adept though they usually are at taxing astronomers' ingenuity.

X-ray satellites have been capable of producing fully two-dimensional pictures of dark clouds. Some dark clouds are studded with X-ray sources, a few of which are certainly identified with T Tauri stars. Occasionally these accidentally discovered ("serendipitous") X-ray objects coincide with visible stars not known in catalogues of T Tauri stars. Optical studies suggest that these stars are very similar to the weaker examples of T Tauri stars but they lack the circumstellar dust shells that characterise bona fide T Tauri objects in the infrared. What relationship could exist between the serendipitous X-ray stars and true T Tauri stars? One suggestion deals with the probable successors to the T Tauri stars. What might an aging T Tauri star look like? Perhaps a less extreme version of a regular T Tauri object? If T Tauri stars abound in dark clouds, shouldn't the stars that they evolve into also be plentiful around these nurseries? How could these hypothetical "post-T Tauri stars" have eluded optical recognition? The primary means of identifying T Tauris is by optical imaging spectroscopy in which photographs sensitive mainly to the strong red emission line of hydrogen are obtained and compared with red or yellow images of the same regions that are not sensitive to this hydrogen line's light. T Tauri stars stand out in the hydrogen line pictures (this is the same prodigiously powerful line that was shown, truncated, in Fig. 5.2 in order to display the plethora of much weaker lines that are also manifestations of the T Tauri phenomenon). Now this technique works successfully only for hydrogen lines above a certain strength compared to the stellar continua. It is very easy to spot a very faint star (faint, that is, in the continuum) with a powerful hydrogen line. However, a brighter stellar continuum, above which is a rather weak hydrogen emission line, is another kettle of fish and these objects might well escape surveys. It is fortunate that their coronæ can be recognised by means of X-ray images or these possible successors to T Tauri stars might have remained unknown for several more years, until optical surveys became more selectively designed to hunt them. It is also relevant to note that the serendipitous X-ray population in the Orion nursery is quite large, leading to the conclusion that there are indeed lots of these weak-lined stars in at least that young association. Perhaps these are the "missing link" between the T Tauri stars and the main sequence.

5.8. Variability

One of the primary challenges to modelling the T Tauri stars lies in their nasty tendency to change unexpectedly. Their optical spectra alter from night to night: emission lines weaken and strengthen at will; hydrogen lines alter in their own way relative to iron or magnesium or sodium lines. Over longer periods (several nights to several months), a T Tauri star, whose spectrum was once full of emission lines, can drop this bright-lined mask abruptly and appear to be a common T Tauri star, with a normal photosphere and very few emission lines. In the ultraviolet, a low-level but constant flickering takes place as if a giant ensemble of solar-like flares were present. However, all T Tauri flares are violent, involving about 1000 times as much energy as solar events. In the X-ray region, several otherwise well-behaved stars have been seen to undergo tremendous variations with flares that last 3 or 4 hours, causing an increase in X-ray emission from the whole star by a factor of 10. Indeed, X-ray pictures of the nursery in Ophiuchus, taken months apart, have been likened to images of a Christmas tree with several independent strings of winking lights! This effect arises because of the strong variability of individual pre-main-sequence stars.

Perhaps the most singular event was the optical behaviour of SY Chamæleontis. Once a slightly variable star, with no obvious regularity, this remarkable object suddenly became a respectable clock, undergoing periodic (once every 6 days) and large optical variations (its blue light output marched up and down over a factor of 6 in brightness). Then, somewhere in the mid-1970s, this pattern vanished and the earlier low-amplitude irregularity reappeared! Presumably the periodic behaviour was caused by a bright (hotter than the photosphere) starspot, carried over the face of the star by a 6-day rotation period. But what a bright spot! And how long-lived and steady!

Radio observations of the T Tauri stars span a far shorter period of time than the optical or ultraviolet monitoring studies. However, in the radio too there have been notable stars. Basically, it has proved to be very difficult to detect radio flux from more than a handful of typical T Tauri stars (as opposed to the rather special objects discussed in Chapter 6.7). Nevertheless, one of the first to be noted was the rather scruffy V410 Tau – an extremely weak T Tauri object, barely admissible to the class. This star, however, revealed by far the strongest radio flux. It was first measured in 1981, seemed only a little fainter by 1982, but had plummetted in radio brightness by mid-1983. It was redetected in late 1983, with a radio signal about 20 to 60 (dependent on frequency) times fainter than its former level! Such wild, dramatic variations in such different spectral regions speak for some kind of intrinsic stellar variability in the T Tauri stars, perhaps in the form of monster flares.

On the other hand, there is another, much gentler, form of variability and one that is extrinsic to the stars. A number of T Tauri stars have been monitored over days from ultraviolet to infrared and, whilst the overall impression is one of a bewildering array of misbehaviours (Fig. 5.15), a small number have revealed definite, correlated changes at all wavelengths (see RW Aur, for example, in this figure). A further crucial element to note here is that the shorter the wavelength, the larger the change in brightness. (See how the ultraviolet swings of RW Aur far exceed the relatively minor changes in the infrared.) This kind of variation has been attributed to eclipses of central T Tauri stars by very large, fluffy, dusty gaseous protoplanets (uncollapsed clouds that could one day form Jupiters). Such dusty obscuration has the wavelength-dependent character observed. This seems an attractive idea but there is a catch. In order to witness any alteration in stellar brightness, you have to look at these young suns from within the planes of their solar systems in which any forming planets will orbit. For one or two stars, it is fine to invoke this mechanism, but how many solar systems could we be lucky enough to view from the right direction? One? Two? Five? Fortunately no-one has claimed that this idea relates to more than a handful of T Tauri stars so it may still be the correct interpretation for these few special cases.

5.9. Magnetic fields

If we believe that T Tauri stars are engaged in the production of massive stellar flares then it is vital to ask about the strengths and distribution of surface magnetic fields on these stars, for we know solar flares to be magnetically orchestrated. One major problem with looking at a distant star is that the spectral signatures of any magnetic fields are inevitably smeared out because we cannot spatially resolve individual portions of active stellar surface. Were we to look only at the entire solar disk, which we know locally

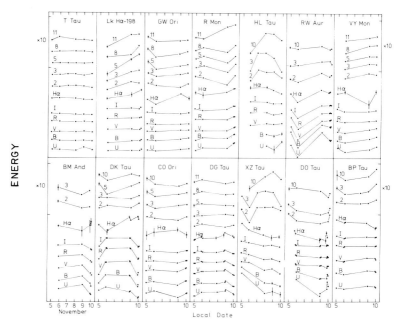

Figure 5.15   A montage of behaviours of different T Tauri stars over a one week period. Presented up the page in each plot are the variations at short wavelengths (UBVRI), in the hydrogen-alpha emission line, and at several infrared wavelengths (labelled by their lengths in micrometres).

to be a crazy quilt of fields, here too we would measure no, or very little, average field strength. What is needed is for some particularly bright or dark spotted region on a T Tauri star to dominate the chromospheric emission or photospheric absorption spectrum, respectively, for long enough for us to detect its magnetic spectral signature. It appears that this may have been achieved very recently for the first time in a T Tauri star, and it shows the average T Tauri fields to be much stronger than the average solar magnetic field. However, we know that the local solar field often exceeds 1000 times the average over the entire solar disk. Consequently, it is likely that these T Tauri stars have fields that are locally comparable with the solar magnetic fields.

It is almost time to turn from the dazzling array of curious observational characteristics of the T Tauri stars to a theoretical view of star formation (Chapter 7). We have met with our best estimate for the nature of the young sun: how did it happen to become the T Tauri star that we think represents the pre-main-sequence sun? How was the perilous journey from cloud to star accomplished? In further preparation for Chapter 7, we will first confront the even more bizarre characteristics of a special group of stars that we have cogent reasons to suspect of being the predecessors of the young T Tauri stars.

6

Nature's womb

And our peace is put in impossible things
Where clashed and thundered unthinkable wings
Round an incredible star
G. K. Chesterton
"The House of Christmas"

Great contest follows, and much learned dust
Involves the combatants
William Cowper
"The Task", book iii

Herbig-Haro Objects – shock waves – interstellar "bullets" – "exciting" stars – molecular bipolar flows – radio images – accretion disks – far-infrared disks – binary stars

6.1. The Herbig-Haro Objects

Of necessity, the earliest phases of stellar evolution take place deep within molecular clouds. Direct optical observations of these processes are therefore impossible but we have recognised the possibility of using an indirect probe of stellar behaviour. Lying in the same dark clouds where stars are born, and apparently associated only with low-mass stars, is a small population of nebulosities, usually tiny, called "Herbig-Haro Objects". The discovery of these was announced in the early 1950s independently by George Herbig, of Lick Observatory in California, and by Guillermo Haro, of the Observatories of Tonantzintla and Tacubaya in Mexico. A picture of the first two Herbig-Haro (abbreviated to HH hereafter) objects appears in Fig. 6.1 which illustrates the knotty character of HH 1 and 2, and the wispy interconnections sometimes found between individual nebular knots or nuclei.

It was, of course, of great interest to determine whether HH nebulæ contained recognisable stars, or were themselves representative of the earliest visible structures of young stars. This exciting prospect was enlivened some years later when a new knot appeared within the HH 2 complex: could this have been a star that had just been born? No starlike condensation could be found, however, even pushing to the near-infrared limits of photographic plates (around 8500 Å). What was rapidly found to be most

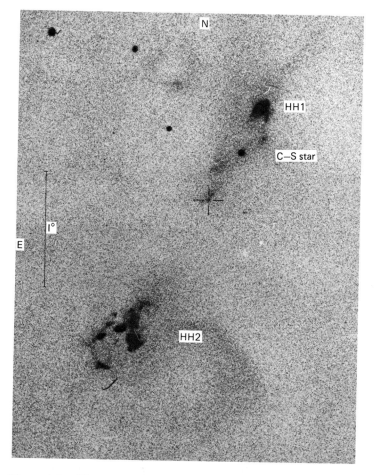

Figure **6.1** The region of HH objects 1 (upper right) and 2 (lower left), showing, in between, the star ("C-S star": the Cohen-Schwartz star) originally thought to have produced them and the location of the radio source announced in 1985 that is probably the true exciting star (small cross). (Photograph courtesy of Burt Jones and George Herbig, Lick Observatory, University of California, Santa Cruz.)

characteristic of HH objects was their spectrum. No continuous emission was typically present, but instead a host of powerful emission lines dominated the spectra (Fig. 6.2). Two theories were advanced by the 1970s to explain these unusual, though not unique, spectral features. One hypothesis attributed the lines to a young star, embedded in the dark cloud, and seen only by being mirrored by dusty dense blobs of gas, which were the HH objects. From the second viewpoint, HH objects were blobs of material either created by a young star, perhaps ejected by it, or were pre-existing cloudlets subsequently struck by a powerful stellar wind. The gas would be heated to a high temperature because of the shock induced by this high-speed encounter between cloudlet and medium. What happens to this hot gas? It has to cool; that's just the first law of thermodynamics: heat flows from hot to cold. Not very profound but definitely useful. So, how can the gas cool; how can it emit energy? It does this by radiating at specific frequencies, appropriate to the atoms in the gas, and it can ditch a tremendous amount of energy in these lines. That idea would explain the powerful emission lines so diagnostic of the HH object phenomenon.

6.2. The continua in Herbig-Haro Objects

Let's clarify some terminology here. We are going to speak of continuous radiation. Recall the spectra of real T Tauri stars (Fig. 5.2). The features that you see pointing up in the diagram, above the average levels, are called "emission lines" because their responsible atoms find themselves at higher temperatures than those that produced the basic background of stellar radiation. Similarly, the features that point down in the diagram represent absorptions by atoms lying in cooler regions than those that caused the background of starlight. This background, together with the absorption lines, arises in the stellar atmosphere. It is this that we shall think of as the continuum, or the continuous radiation from the star. We will be using the term continuum solely to distinguish the starlight component of radiation from the emission-line component in the discussion below. (Strictly, "continuous" connotes the absence of any features, in emission or absorption, but

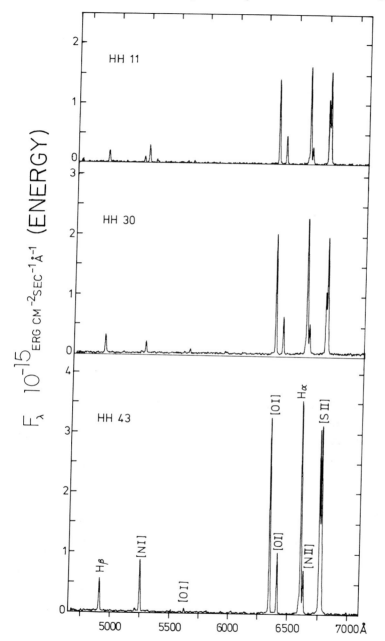

Figure 6.2 Optical spectra of three HH objects showing their powerful emission lines. Wavelengths are in Ångstroms.

we are relaxing this terminology to include all radiation that derives from the star.)

Back to the HH objects. In the reflection scheme, the emission lines should be significantly polarised since the supposed starlight that we can observe has its electric field in a specific direction, relative to the geometry of the HH mirror and our viewpoint. In the shock cooling picture no appreciable polarisation of the lines would be expected since the lines arise precisely where the gas is hot, and that's at the HH object itself. So no indirect viewing is involved. Whenever this polarimetric experiment has been carried out on an HH nebula, it has demonstrated the relevance of both therories! True, the emission lines are quite unpolarised, which shows that they have their origin at the location of an HH object and not in some displaced star. However, the continuous emission which, although not a general characteristic, does accompany some HH objects, is always found to be polarised, occasionally very strongly. Consequently, HH objects play a dual role: their gaseous component seems to indicate the presence of a high temperature region that cools after it has suffered a shock, in a collision; their dusty component (mixed in with the gas) mirrors the starlight from a young star that in some manner must be responsible for the production of the HH cloudlet, or the shock, or both. This work represents a rare example of a critical observation that left everybody content with their theories!

The total population of bona fide HH nebulosities of which we are currently aware numbers only about 100. Not all HH objects have the same spectrum, as Fig. 6.2 shows. They all have strong atomic emission lines but there are subtle, or sometimes substantial, differences in the relative strengths of these signatures. Of course the prominence of the reflected starlight component (the continuous spectrum below the lines) is a function of the distance from an individual HH nebula to its responsible young star, the local fogginess of the intervening interstellar medium, the amount of reflecting dust particles mixed into the HH object's gas. So the brightness of the continuum does not tell us anything unique about the environment of the HH nebula. What about the lines?

6.3. Shock wave models

Theoretical models have been calculated that represent the physical consequences of a rapid collision between a gaseous blob and the interstellar medium. The different calculations cover a wide range of relative velocities of impact between cloudlet and medium, and differing initial densities for the ambient medium. Variations of these parameters produce a set of models that indicates the different temperatures to which the shocked gas is heated. We have stated that the emission lines in an HH object serve the purpose of allowing the hot gas to cool by carrying away energy through radiation. The lines are such critical coolants for the gas that it should be no surprise to learn that the intensities of specific lines are diagnostic òf the circumstances of the shock; in particular, the velocity and how "hard" (dense) the wall of the medium is into which these blobs plough. To provide some scale for these processes, how fast is a typical HH colliding with the dark cloudy medium? About 100 km/sec or 230,000 m.p.h.!

Notice that we used the term "relative velocity". It makes no difference to the emergent line spectrum whether an ejected blob cannons into a stationary medium, or a high speed wind picks up and accelerates an ambient, stationary blob. Only the relative velocity – the difference in speed between blob and gassy medium – is important. But which of these two possibilities is correct, in detail? This is still controversial although there have been some fairly surprising discoveries about HH objects that offer us a clue.

6.4. Proper motions

If we observe any celestial object and ask about its velocity, its motion in the sky, then there are two distinct motions about which we can talk. The first is related to the radial velocity, namely the component of velocity that occurs along the line of sight (along the "radius vector" joining observer and object). This radial motion, you will remember, is measured by the Doppler shift of atomic or other spectral features. The second form of motion arises within the plane of the sky rather than perpendicular to it. Since the earth is trundling around the sun, and the sun around the Galaxy, you can never trust radial motions to be entirely a measure of true celestial velocities. You always have to specify a frame with respect to which you will quote a radial motion. For example, you measure "geocentric" velocities because you are on the surface of the earth. You can convert these to motions with reference to the centre of the sun ("heliocentric motions") after allowing for the motion of the earth in its orbit around the sun. You can even refer to a frame whizzing around the Galaxy along with the sun and its nearest neighbours – the so-called "Local Standard of Rest". However, if you observe a modestly distant object you will find that it also has a motion all of its own across the sky, a motion "proper" to that object. So proper motions denote the second component of velocities and these motions are unaffected by the Doppler shift (which is produced only by line-of-sight velocities).

As we noted in Chapter 5.3, scanning dark clouds at infrared wavelengths offers a way of penetrating much more deeply into the murky stellar nurseries than does light. The region surrounding HH 1 and 2 has been observed in the near-infrared part of the spectrum with the intention of seeking potential embedded young stars that might be responsible for the production of these knotty complexes of shocked nuclei. Only one infrared source was found, lying along the line connecting HH 1 and 2, and coincident (Fig. 6.1) with a faint visible star. The spectrum of this star showed it to be of T Tauri type, albeit a mild representative of the genre. It was, therefore, tentatively identified as the potential parent star for both HH nebulæ, especially in view of its suggestive alignment with these. Herbig, fascinated by the behaviour and structural variations in these HH objects, had acquired a long series of photographic plates spanning some thirty-four years. These enabled astronomers at the Lick Observatory to determine the motions peculiar to the individual knots in both nebulous complexes. Their findings were startling, as Fig. 6.3 reveals. Each knot of HH 1 is moving away to the NNE, and each component measured in HH 2 is travelling away, but in the opposite direction! These data are strongly suggestive of physical motions, perhaps due to ejection of the HH knots from some star that lies between them. How fast are these veritable bullets moving? At speeds of the order of 250-300 km/sec, or almost 700,000 m.p.h. across the sky!

The chase was now on! The Lick astrometrists (astronomers who specialise in the very accurate determination of the positions of celestial sources, from which the motions in the plane of the sky

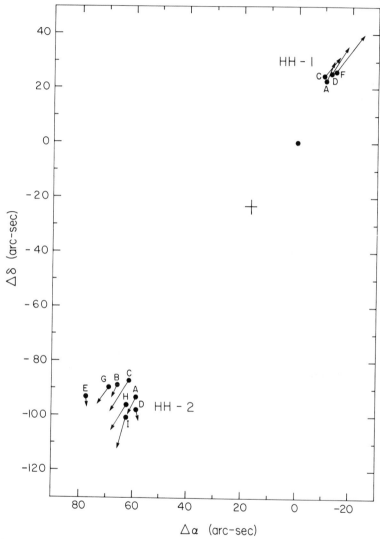

Figure 6.3 Proper motions of individual knots in HH1 and HH2 away from the radio source that is likely to represent their exciting star (the small cross). Axes represent the components of proper motion in equatorial celestial coordinates; the x-axis is ''Right Ascension'' (delta alpha); the y-axis is ''Declination'' (delta delta).

can be measured) began to measure photographs of several other families of HH nebulosities, hoping to confirm this remarkable behaviour in other systems. Their efforts met with great success. Although not every HH object can be demonstrated to be in rapid motion, many complexes include one or two knots whose motions across the sky are sufficiently fast to be recognised and determined with some precision over a baseline of only a few years.

It was not until the spring of 1985 that a deep VLA (Chapter 6.7) radio map of this region of the sky revealed another potential candidate for exciting HH 1 and 2. The new candidate lies symmetrically between HH 1 and 2, along the line joining them. It has an optical counterpart in the form of a faint red conical nebula pointing towards HH 2 (see Fig. 6.1), but is very faint at short infrared wavelengths, becoming highly conspicuous only in the far-infrared. The location and radio properties of this new source earmark it as the true exciting star of these HH nebulæ and suggest that the weak T Tauri star was an "infrared herring!".

6.5. How to find hidden exciting stars

Interesting in their own right though HH nebulæ are, let us not lose sight of our reason for introducing them. Suppose you take a typical, small dark cloud containing a visible HH object. No obvious infrared source has been found within the cloud even with ground-based infrared techniques. What does the HH object signify, and how can we use it to learn about the earliest phases of star formation? By good fortune we find that our HH nebula is in rapid motion across the dark cloud and we have defined the direction of its path. Even though we do not understand the physical process responsible for the motion, we can recognise, at least intuitively, that there must be a causal connection between something in the cloud and the racing HH nebula. Somewhere along the backtracked path of this blob must lie a star, of some kind, that may have ejected it or at least have sent it on its way by some indirect means. How do we find it?

The earlier the phase of evolution of a star, the fluffier it will be. Fluffiness sounds like a singularly unphysical concept but it connotes a useful qualitative measure of the degree of collapse of a cloud fragment. A fragment that has recently begun to collapse will still be cloudlike – diffuse; of low density more or less everywhere. Later, the protostellar core will be established – a tightly bound (very negative potential energy: remember Chapter 2), dense condensation within the central part of the cloud. So, fluffy stars

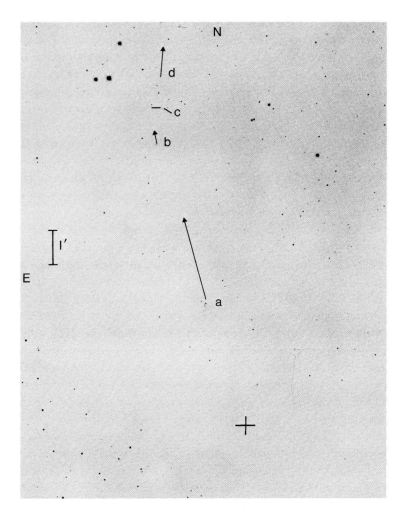

Figure **6.4**   The pattern of proper motion vectors for several HH knots (a-d) in a family that led to the discovery of the location of the hidden exciting star (cross) within the dark cloud that crosses this photograph diagonally from left centre to bottom right.

are ones that have given up very little of the initial potential energy of their parent clouds. What happens when this potential energy is leaked away, enabling further collapse of the core? The temperature of the dusty component of the cloud goes up, producing infrared radiation to balance the lost gravitational potential energy.

So what's wrong with our hypothetical cloud that seems to contain no central near-infrared object? Nothing, but we're too early; too early for adequate heating of the dust to a temperature detectable by ground-based telescopes at these short infrared wavelengths. Instead of a thousand or even a few hundred degrees Kelvin, our fluffy protostellar core has warmed its environs to only a chilly few tens of degrees. At what wavelength would this dust radiate? In the far-infrared which necessitates observations from aircraft, balloons or even satellites. Therefore, armed solely with the location of the HH object and the direction of its motion, we would undertake an airborne survey of the cloud along the wake of the errant nebulosity.

Whenever this has been attempted, it has been remarkably successful. In fact Fig. 6.4 illustrates one rather exciting situation in which the infrared source responsible for an entire family of related, moving knots was located. The cross marks the location of the optically-hidden star and you can see just how dark and inscrutable is the empty face of this dense cloud that conceals the young star. This kind of demonstration, based purely on the morphology of a situation, shows the predictive power of the hypothesis that some stars have ejected blobs of gas like bullets in the interstellar medium. Predictions lend weight to hypotheses, even in the absence of a fully-fledged physical theory of what is really taking place. This particular hypothesis tells us that we can trace back the lineage of stars well beyond their epoch of visibility, closer and closer to the moment of initial establishment of the protostellar core itself. Indeed, one clear consequence of the recent highly productive Infrared Astronomical Satellite is that we can undertake an examination of all dark clouds in the sky, (a very laborious task from airborne observatories) looking for evidence that embedded objects have taken their first steps along the road to star formation. This we can do simply by looking for very cool far-infrared sources associated with individual clouds. Later we can seek other activity represented perhaps by extremely faint wisps of HH nebulosity about which we don't yet know. Just how young are these far-infrared emitting sources? No-one knows for certain but, in one picture, they are certainly younger than 100,000 years, and could even be a mere few tens of thousands of years old!

In a few cases we are fortunate enough to witness both sides of this twin-ejection mechanism for HH objects, for example the HH 1 and 2 system. What happens if these projectiles are sent straight into the body of the parent dark cloud in only one direction? Then we would see only a one-sided phenomenon. Statistically we expect this to be seen much more frequently than the two-sided situation. Therefore, we'd expect to see HH objects mostly approaching us (blue-shifted) and not their receding (red-shifted) counterparts, if there were any. How do we find out whether the star has been active in both directions when we can only find visible evidence in a single one?

6.6. Molecular probes of dark clouds

To solve this problem of obscuration by the cloud we focus our attention on a different probe now; on carbon monoxide (CO) molecular radio emission. The microwave emission from this abundant constituent of dark clouds is readily detectable.

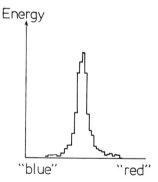

Figure **6.5** Typical CO line profile of a dark cloud. Plotted on the x-axis is frequency, which is equivalent to velocity (via the Doppler shift). Points left and right of the central peak represent gas motions towards and away from us, respectively.

However, our concern is not with the strong spike of emission (Fig. 6.5) so common in clouds but with the potential existence of any perturbations to this spike that could tell us about localised, high-velocity flows of material, such as might be stirred up by an active young star. Suppose we had such a star, buried deep within a dark cloud, ejecting material from the vicinity of its surface by whatever process leads to the bidirectional structure of HH flows.

Then this engine would mechanically push out material not physically associated with our active star, simply because there's no place else for this material to go. This "snow plough" effect is really nothing more than a consequence of the principle of conservation of momentum. A tiny bullet, of minuscule mass, travelling at enormous velocity, will shove along a much larger mass of ambient medium but this swept-up mass will have a proportionately smaller velocity. (It's the product of mass and velocity, which product we call "momentum" or "linear momentum" to distinguish it from angular momentum [Chapter 5.6] that is conserved in all directions.) By analogy then, our optically invisible star will jostle the gas of the dark cloud, driving a bidirectional flow of cloud matter away from it. We'd predict that such an active star should be sitting between two differently Doppler-shifted (Chapter 3.2) lobes of gas. This is called a "bipolar flow" and the discovery of these flows has proved of great value in defining the frequency and energetic requirements of this bizarre form of early stellar activity.

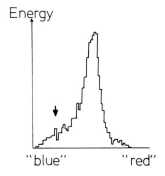

Figure **6.6**   A high-velocity wing to a CO profile. Note the extended tail of emission to the blueward side of the peak (the wing's centre is shown by the arrow).

There is a remarkable keystone object called "L1551 IRS5" which means only that it was the fifth near-infrared object to be found during a survey of the 1551st visibly-known dark cloud out of an optical catalogue by the astronomer Lynds. Microwave observations of L1551 revealed CO profiles that showed much more than merely the central cloud emission profile: at some locations we see a weak, red-shifted velocity component; at others, a blue-shifted one (Fig. 6.6). If we assemble all these observations of weak high-velocity wings taken at different places in the cloud we can map the separate spatial distributions of the blue- and red-shifted molecular gas. This map (Fig. 6.7) strikingly illustrates a bipolar flow: on either side of the infrared source IRS5 there is an entire lobe of material, systematically moving either towards us (the south-west lobe) or away from us (the north-east mass). (Incidentally, if you're puzzled about the apparent reversal of east and west in astronomical pictures of the sky, just think about holding a terrestrial map over your head, pointing north to the celestial north direction. Notice that east and west are interchanged!) In the remaining two quadrants there are no high-velocity wings seen at all, only the strong central cloud profile.

Let's examine these lobes more closely. Nothing can be seen in, or through, the dark cloud that coincides with, or overlies, the red-shifted CO lobe. The systematic motions of the molecular gas take place deep inside the cloud, without any visible consequences. But the blue lobe is really intriguing: inside it we find no less than two rather distant, well-defined, bright, high-speed HH nebulæ, flying away from IRS5 – visible tracers of the directed activity of this remarkable object. These rapidly-moving knots are emerging from the surface of the dark cloud which is the only reason that we can photograph them. But we can see that the HH nebulæ represent only a small portion of a really rapid stream of energy that has ploughed up the local interstellar medium. The HH objects are moving at around 300 km/sec while the swept up, much more massive, molecular lobes are travelling only at 20 km/sec. Here we see conservation of momentum in action.

6.7. Radio studies with the "Very Large Array"

Now the interest switches to another, longer-wavelength part of the radio spectrum. The Very Large Array radio telescope was brought to bear on L1551 and, on the first attempt, not only was IRS5 itself detected but the radio-emitting region was spatially resolved. This volume of ionised gas has an elongated structure (Fig. 6.8) whose long axis aligns with the axis of the much larger and more distant CO lobes! All the way from the immediate vicinity of IRS5 to far out into the dark cloud, this amazing little powerhouse maintains its memory of the direction in which it has ejected material – across a range in distance of over a factor of 100 from the star.

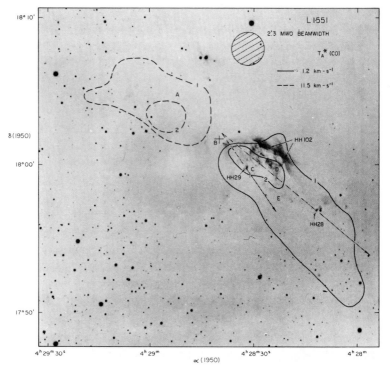

Figure 6.7   A CO bipolar flow in the Lynds 1551 dark cloud. All solid contours represent approaching molecular gas; all dashed, receding gas along our line of sight. The centre of this flow is close to the cross at the letter B where the infrared object L1551 IRS5 lies. In the blue CO flow note the high-velocity HH objects 29 and 28, flying away from IRS5.

When optical astronomers observe stars from the surface of the earth, they do not see points of light. What they do see is a small, bright patch which represents the blurring effects of the terrestrial atmosphere on the light rays that come to us from the star. Suppose the telescope were lofted into orbit, as NASA's 2-metre reflecting Space Telescope will be, in 1989. What do star images look like from space, freed of the fuzziness caused by an atmosphere? Stars will definitely appear smaller but still won't be points. The limit on sharpness, however, is now set by theory that describes the ultimate optical performance of the telescope. A process called diffraction actually enables light rays to travel in other than straight lines sometimes. Precise points of light, when viewed through even the most perfect telescopes, are rendered as tiny circular patches

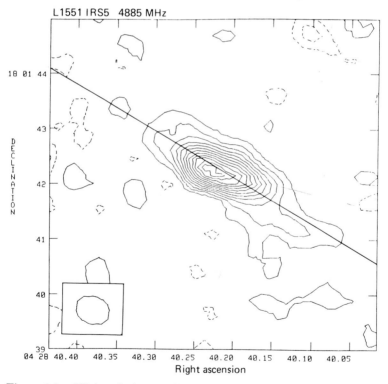

Figure **6.8**   VLA radio image of L1551 IRS5 showing the elongated flow of ionised gas. Solid line represents the axis of the much more distant CO flow in Fig. 6.7. Insert shows radio beam size.

surrounded by fainter, concentric rings of diminishing brightness. The diameter of these central patches is proportional to the wavelength of light used and inversely proportional to the mirror size: larger mirrors yield smaller patches, which is why astronomers are always clamouring for bigger, hence better, telescopes. It is also true that an optical telescope yields fuzzier images when used at infrared wavelengths, since these are longer than normal visible wavelengths.

Diffraction always imposes the ultimate limit on the fineness of spatial resolution of a telescope, no matter what the part of the spectrum in which it is to be used. Radio wavelengths are enormous (you could pick them up and hold them!) compared with those of light waves, hence the large diameters of radio antennæ. Great ingenuity has gone into the construction of arrays of antennæ which can yield resolutions appropriate to the greatest separation of a pair of antennæ, by a technique called "aperture synthesis". You need only have dishes 30 kilometres apart to achieve the effective resolution of an entire antenna, 30 km in diameter, although, of course, you still only receive energy from space at a rate appropriate to the actual collecting area used. The Very Large Array is such an instrument. Remarkable though it is, the VLA still can't resolve the thickness of the elongated image of L1551 IRS5. The length of IRS5's radio image is no problem but the width is still in accord with the diffraction patch (Fig. 6.8). Calculations, however, suggest that IRS5 is sitting in the middle of an extremely well-collimated radio-emitting region – a bidirectional "jet" whose length could be over 100 times its cross-sectional width! There is thus a striking structural similarity between this twin jet and those associated with those most perplexing of astronomical powerhouses, the quasars. This is not to say that young stars and quasars operate on the same principles, and certainly does not imply that a forming star contains a quasar! But there is something ubiquitous and robust about the phenomenon of jets. Ideas and models that can describe jets from young stars are still at the cutting edge of theory and it would be beyond the scope of this book to attempt a detailed discussion. However, there are some broad aspects of this work that we can explore.

6.8. Accretion disks

The only sensible connection between objects as diverse as the strongest celestial engines we know of and low-mass protostars must be rotation. On many different scales in astrophysics, astronomers have proposed disks of material, rotating around central energy sources. Generally these are termed "accretion disks", which implies that the ultimate fate of much of the material in one of these disks is to fall onto, and be eaten by, whatever object lurks at the centre of the disk. Accretion disks are found around black holes – those ravenous maws that symbolise the end of the life of a high-mass star. Around the turmoil at the nuclei of many galaxies and the central engines of quasars accretion disks crop up again. For our purposes, too, these disks will occur around protostars and much of their material will spiral slowly into the forming stars. Where do the disks come from? Our protostar is created as a denser lump at the middle of what is basically a collapsing spherical cloud of gas and dust. During this collapse, momentum must be conserved – not just the linear momentum of bullets colliding with gas blobs, but angular momentum too (Chapter 5.6).

Back to our forming star. During the collapse, any material from the original cloud that has low angular momentum must fall right into the star or into its immediate vicinity. However, what about matter with initially high angular momentum in the cloud fragment? This cannot collapse into the core: to preserve its momentum it must stay far from the axis of rotation. This creates a disk in which all high-spin matter can reside. Exactly what happens to that matter we shall return to, in more detail, in Chapter 10. For the moment we note only that the collapse of a rotating cloud fragment must inevitably create a disk. Now this disk can be very large at first, much larger than our known solar system, but various physical processes can leach away the angular momentum (this doesn't violate the conservation principle since the application of external forces obviates the need for this conservation – it operates only in a system that has no external force), driving material closer and closer to the central star; eventually, perhaps, right to the stellar surface.

6.9. Have we seen the disks?

Very recent airborne far-infrared observations of the sources that represent HH-exciting stars have yielded some remarkable results. From the pattern of HH knots or proper motions we can readily infer the direction in which material flows from an exciting star. Fig. 6.9 shows two examples of this. The striking chain of HH nebulæ, HH 7-8-9-10-11, represents what could be a repeated (almost regularly too) series of ejections of HH objects from the site of the embedded star (the cross in Fig. 6.9a). The axis of the HH chain must be the flow direction. Likewise,

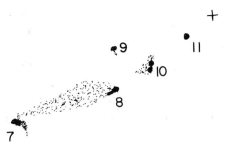

Figure **6.9 a)**   Sketch of the disposition of the unusual chain of HH objects 7-11 and the location of their dust-embedded exciting star SVS13 (cross).

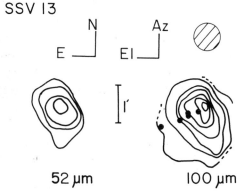

Figure **6.9 b)**   The far-infrared maps of SVS13 showing real structure perpendicular to the axis of the HH objects. Infrared beam size is also shown, as are the "Az" (azimuth) and "El" (elevation) directions for the altazimuth airborne telescope.

Fig. 6.10b presents a photograph of a star that rather surprisingly popped up a few years ago where none had been seen previously, right in the middle of a loopy, elongated nebula containing one little HH knot (HH 57A). The structure of this nebula leaves no doubt that matter has flowed in a roughly north-south direction from this star. Now let's examine far-infrared images of these stars

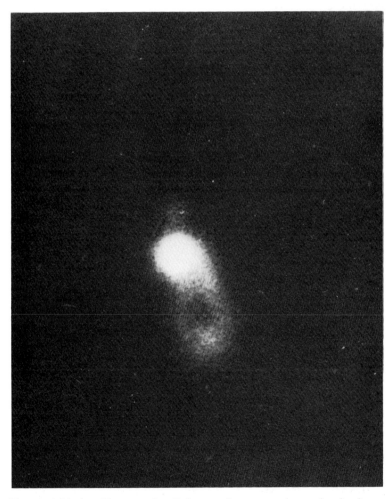

Figure **6.10 a)** Photograph of the newly-appeared star in the loopy HH57 nebulosity. (Courtesy John Graham, Cerro Telolo Inter-American Observatory, Chile.)

(at a wavelength 200 times that of yellow light). In Figs. 6.9b and 6.10b we see the far-infrared contour maps for the exciting stars hidden at the end of the HH 7-11 chain, and for the now-visible star in HH 57. Also shown for direct comparison is the beam size used for these airborne observations.

Parallel to the flow directions in both cases the contour maps are narrow: only the expected diffraction size is detected. But perpendicular to these flows we clearly see how much more elongated are the dust structures that radiate at these very long wavelengths. In short, we have spatially resolved something perpendicular to the flows but can see no evidence of extension along the flow axes. In what follows, and in Chapters 7 and 10, we shall often refer to "disks". There truly should be disks around young stars – with the connotation of dinner plates. However, even earlier there will be flattened infalling clouds of gas and dust, more like round pillows than plates. We shall continue to use the term "disk" but your mental image should depend upon the stage of evolution implied; for example, disks (plates) around T Tauri stars

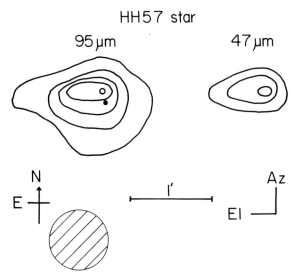

HH57 star

95 µm   47 µm

Figure **6.10 b)**   Far-infrared contour maps of the source at the stellar position (open circle). "Az" and "El" are as in Fig. 6.9b. The HH nebulous blob below the star in a) is represented by a filled circle in b). Again, the extended nature of this source, in the east-west direction, is obvious.

but disks (pillows) around the HH-exciting stars described in the present section. These remarkable maps are our first direct views of giant circumstellar dusty disks around the HH-exciting stars (or for that matter around any young stars)! Their scale is expansive. Typical radii (for these two and several other examples) are several thousand AU – far wider than our known planetary system, although certainly comparable with the size of the debris cloud hypothesised to surround our solar system, which is the home-base of most comets.

6.10. The binary star L1551 IRS5

Finally, an even more startling image is Fig. 6.11, which shows a high-frequency VLA radio map of L1551 IRS5. We see a double object (the solid contours) – apparently a pair of protostars (that mutually orbit one another in a "binary system")! The brighter of these two high-frequency objects sits right at the centre of the lower frequency structure (the dotted contours) that represents the two-sided radio flow (Fig. 6.8) or "jet". The inner contours of the low-frequency radio jet seem to emerge perpendicular to the projected orbital plane of this putative binary!

Could all HH-exciting stars be members of binary systems in which the jets somehow tap the rotational energy of orbital motion (see Chapter 10.4)? It is too soon to justify this conclusion but the field is really jumping again, with all the new data coming from a giant international effort to observe IRS5 by every available technique. This is where we must leave the stars of HH objects but our discussion has taken us right onto the frontier of knowledge about these fascinating nebulæ.

6.11. What are the HH-exciting stars?

Can we definitely establish a connection between the stars (often optically invisible) that excite HH objects and the T Tauri stars? In a few cases, the HH-exciting stars are visible and can be recognised as T Tauri objects, usually with very strong emission-line spectra (perhaps the "most active" of these stars?). It is important to remember that the HH phenomenon does not last for ever; in fact, the shock-excited gas can cool through its powerful optical lines in about one to ten thousand years. Consequently, only the most recently-formed HH nebulæ can be detected. The vast body of T Tauri stars is generally not associated with HH nebulæ. Very few have photographically-detected HH nebulæ, and a few more show spectroscopic evidence of HH-like emission lines but are not nebulous in appearance, perhaps connoting the very recent creation of an HH object. We have to conclude that only the very youngest T Tauri stars make HH nebulæ. Likewise, radio continuum surveys of T Tauri stars and HH-exciting stars lead to the same conclusion: the highly-directed rapid mass flows, responsible for radio jets and related to HH objects, are not characteristic of older, more readily visible T Tauri stars.

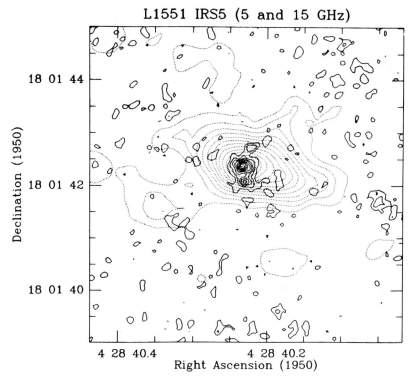

L1551 IRS5 (5 and 15 GHz)

Declination (1950)

18 01 44

18 01 42

18 01 40

Right Ascension (1950)

4 28 40.4   4 28 40.2

Figure **6.11**   New VLA images of L1551 IRS5 that overlay low-frequency (dotted contours) and high-frequency (solid contours) maps. At high frequencies, IRS5 seems to consist of two objects, one above the other in the map.

From our infrared observations at the longest wavelengths we can establish the bolometric luminosities of the HH-exciting stars. With a single exception, all are very much less than 250 solar luminosities. We are certainly not dealing with a phenomenon of high-mass stars. In fact, all (but two) of the exciting stars could be interpreted as protostars, still accreting from their parent clouds, that will one day wind up as roughly solar mass pre-main-sequence stars. The conclusion, therefore, is that protostellar objects now excite HH objects and that these protostars are the precursors of visible T Tauri stars.

Now it is appropriate for us to try to synthesise a theoretical sense of how stars are born from the darkness of interstellar clouds, and to understand what mechanisms might drive the zany behaviour described in Chapters 5 and 6.

7

How and why clouds collapse

There are many events in the womb of time
which will be delivered.
William Shakespeare
"Othello"

Choose a firm cloud, before it fall
Alexander Pope
To Mrs. M. Blount

*Clouds and stars – answers to the "cosmic questionnaire" –
origin of cloud complexes – instabilities – forces on a cloud –
role of magnetic fields – accretion – from cloud to star*

7.1. Some stars have formed recently

The Galaxy is old: older than its oldest stars which have
been dated at 12 thousand million years. We have seen that the
hottest stars expend their nuclear fuel so rapidly that even the
greatest portion of their lives, that spent on the main sequence,
cannot exceed 2 or 3 million years. Therefore, these stars at least
were formed recently. In Chapter 5 we met the T Tauri stars –
much more abundant than the hottest stars and consequently of
lower mass (remember the mass spectrum and birth function of
stars: Chapter 2.10). These curious stars seem to evince signs of
youth, from their erratic and rapidly variable behaviour to their
obviously intimate association with dark clouds. The T Tauri stars
cannot be so old as the Galaxy so they too must have been formed
relatively recently. We have identified the dark clouds as the likely
nurseries for newly-forming stars. However, observations,
particularly infrared and radio molecular ones, tell us that any old
piece of a molecular cloud complex is not a suitable location for star
formation to take place. Rather we must seek the "cores" – the

densest portions – to find evidence of recent stellar births. What is the "missing link" between clouds and stars? This question will occupy Chapter 7 and we shall find it useful to consider that perhaps high-mass and low-mass stars form by quite different mechanisms.

7.2. Answers to the "cosmic questionnaire"

In Chapter 5 we described the fascinating properties of T Tauri stars and we were able to construct an HR diagram for the entire Taurus-Auriga nursery. From that HR diagram, and its interpretative overlay (Fig. 5.9), we can deduce the basic parameters of these stars. It is, therefore, appropriate that we now attempt to answer the various issues raised by the "Cosmic Questionnaire" posed in Chapter 4.7; for convenience we use the same numbers for answers as for the questions.

1 and 4. The observed T Tauri stars show us that they represent the results of star formation in the range from about 0.2 to 3 solar masses. In clouds that contain T Tauri stars we often find slightly higher-mass, more luminous, warmer young stars, in the approximate range 3 to 15 solar masses. Above that, our attention usually switches to more distant nurseries. In Orion's Great Nebula we can see young (they're always young, even when they are evolved and about to die, of course! Chapter 2.9) O-stars; brilliant beacons with upwards of 20 solar masses. Far beyond Orion's 1500 LY we can see the giant O-associations in Casseiopeia, 7000 LY distant; clusters of forming, and formed, luminous O-stars, making ionised gas clouds, puffing winds into dark globules, and generally showing off as one would expect the most massive stars in the Galaxy to do. Could there be T Tauri stars mingled with these O-stars? In Orion there are; we see them readily enough with modest to large telescopes; in the big O-associations they would be much too faint optically to have been detected and recognised as T Tauris, but they could exist. We do see evidence, then, for formation of stars from the lightest to the heaviest though not always in a single nursery.

2. The vital statistics for T Tauri stars are as follows: typical stars are around 1 solar mass, with surface temperatures of 4000 K, radii

3 times the sun's, and ages anywhere from 100,000 up to 20 million years. This spread in stellar ages hints that star formation does not have to take place simultaneously in all parts of a single nursery.

3. Chapter 5 discussed the nature of the X-ray, optical, ultraviolet and infrared emission from T Tauri stars. For the hotter young stars, little changes in our story. Radio emission attends visible low-mass young stars very infrequently and giant flarelike events seem implicated. Deeply embedded high-mass stars are often recognisable by radio continuum emission, not of the stars themselves but of the associated surrounding ionised hydrogen regions.

5. A good way to decide upon the location of the dominant extinction due to dust in T Tauri stars is to compare the obscuration of pairs of adjacent stars. Were the dust to be principally intracluster in nature, the obscurations along these adjacent lines of sight should be very similar. By contrast, were circumstellar shells more important, the pairs of stars should have totally unrelated values of obscuration. In fact, we find in favour of general intracluster reddening, though a few specially-selected stars clearly suffer monstrous circumstellar extinction. We met some of these selectively dimmed stars in Chapters 6 and we'll encounter them again, in Chapter 10.

6. Star formation, for clouds creating preferentially low-mass stars, seems rarely to attain 10% efficiency. Even for nurseries like Orion, where T-associations are mingled with O-associations, the star-forming efficiency lies in the range 1-10%. There are claims that some dark clouds (one notable lies in the constellation of Ophiuchus) have achieved as high as 35-50% efficiency on the basis of their invisible (except at infrared wavelengths) population of presumed main sequence stars. This is a controversial issue for the assignment of masses to objects known only from their infrared properties is fraught with uncertainties. But it may well be that these claims are accurate, in which case it would seem that the particular environment of a cloud can influence its global star-forming ability.

7. It is to Taurus-Auriga that we should turn to answer questions about the mass spectra of young stars for, at only 500 LY distance,

we can survey this nursery with minimal bias against very faint (very low-mass) objects. Taurus-Auriga's mass spectrum is very similar to the birth function at least up to 3 solar masses. Beyond that we probably face severe problems of observational bias for, in order to sample higher-mass stars, we must search more distant nurseries and run the risk of missing their low-luminosity (low-mass) components. It is hard, however, to avoid the feeling that stars of the highest masses may form only from the giant molecular complexes (Chapter 3.7), close to the spiral arms and under the recent influence of the Galactic density wave, whereas our sociable T Tauri stars may be capable of forming in almost any cloud of adequate mass (at least a few tens of solar masses). Therefore, high-mass stars may require other circumstances for their generation (see Chapter 9, for example).

8. This question formed much of the substance of Chapter 5.

9. Chapter 10 will deal with these more parochial aspects of star formation.

10. Chapter 6 explored in great detail the interactions between young stars and their parental clouds. It turns out that the dominant issue, insofar as it may affect future star-forming activity, is simply whether all the vigorous churning up of clouds by active young stars could provide adequate turbulent pressure support for the cloud. If so, perhaps other potential cores would be inhibited from collapsing and the subsequent formation of stars. It is certainly plausible to consider this source of support, especially if star formation rates are brisk among the stars that are the focus of Chapter 6 and that genuinely disturb their surroundings. But at issue, really, is the number of these stars that we can expect to populate a cloud. Observations illustrate, too, that even where these stars have patently been stirring up the cloudy medium, one finds pieces of dark clouds on the verge of gravitational collapse (see Chapter 7.5). Clearly, star formation in these clouds is a reasonably robust process and other low-mass stars do not seem to influence adversely the chances of forming new ones of their kind.

7.3. The origin of giant cloud complexes

Perhaps we should sit back for a moment and take the "big view" before we become too involved in the intricate details of individual collapsing stars. We have decided (Chapter 4.3) that most modern star-forming activity in the Galaxy occurs within the giant molecular complexes – vast, dark, fertile oceans of gas, dust, molecules – fertile with the next generations of stars. Suppose we ask what are the origins of these giant complexes, since they are a prerequisite of young stars. One has the impression that, given one of these complexes, the emergence of stars is assured; it is just a matter of being patient for about ten million years. Further, we find observationally that once the high-mass, luminous O-stars appear, the destruction of their attendant clouds (or clumps) is also inevitable within another thirty million years or so.

So, where do these all-important complexes come from? Why do they congregate along vast spiral structures so many thousands of light years in scale? Finally, why are they gregarious; that is, why are complexes found near other complexes in titanic assemblages accounting for tens of millions of solar masses of material? Density waves provide at least some of the answers. It is their overall, Galaxywide organising ability that dictates the location of giant complexes along spiral arms. The effect of these density waves is to induce shocks in the Galactic gas disk; behind the shocks (where the density wave just passed) the interstellar medium is compressed. The compression leads to condensation into clouds, and organisation of these on the large scale occurs (but not the formation of individual stars yet). It was originally something of a mystery how star formation could arise in such a narrow region as photographs of spiral galaxies implied. So-called "grand design spiral" patterns were revealed – very sharply-defined arms (generally two-armed spirals) – that spoke for the rapid initiation of star formation. Since the gas involved in spiral waves was known to spend only a very short time in the arm, this rapidity of mechanism was a problem. The problem was subsequently solved by the idea that a shock developed behind the arm in what could be a very narrow region, triggering eventual star formation.

It will be helpful, here, to discuss what we mean by "instability". Imagine a pendulum bob, hanging from your ceiling on a long chain. Kick the bob and see what happens. The pendulum will jump crazily a few times, and then settle down to regular oscillations that get smaller and smaller, eventually dying away all together. We say that the pendulum is "stable against" being kicked. What's the alternative? It could be "unstable to" the same kick. If it were, it would be a very strange pendulum! It would show this strangeness by swinging in ever-increasing arcs, leaping about until eventually it broke the chain and flew away. This sense of a wildly-growing response to a small stimulus is connoted by instability, as opposed to stability where all response eventually decreases and the initial, unperturbed state is recovered.

A cloud can be "unstable against gravitational collapse" if its response to a tiny contraction in size (or a push on its outside) is to shrink in a runaway and accelerating fashion until it is entirely compressed near its core. Many everyday situations are inherently stable, fortunately for us! In a moment we shall meet an instability that depends upon an interaction between gravity and magnetic fields. The end-product will be that blobs of gas (soon to be cloud complexes) cannot help sinking into "pockets" in the magnetic field. In Chapter 7.6 we'll encounter a protostar that is perfectly happy with an existing physical situation but, when one aspect of this situation changes, the star will become unstable. Not every instability has to be one against gravity: this latter case is an instability against convection. Initially the star is heated, most unusually, from its outer layers. This gradient in heating establishes a situation in which any slight convective bubbling inside the star would rapidly diminish and could not spread to the rest of the star. Later, the heating switches to the stellar core, setting up a very different physical environment in which slight convective motions inside will amplify, and swell to involve the entire star.

To return to our cloud complexes, it is further true that the shock-compressed gas can be gravitationally unstable to collapse; that is, the mutual attractions of one piece of gas on another act to create a runaway contraction. Calculations of the scale of this instability suggest that one can get between 10 and 100 million solar masses of material to collapse in chunks of order 1 kpc (3000

LY) in length. This idea would very naturally provide us with complexes of cloud complexes.

To make individual complexes requires another theoretically-demonstrated instability involving the magnetic field and the Galactic gas. This one explains that gas always tends to concentrate into "pockets" suspended in the Galactic magnetic field (Fig. 7.1), leading to blobs between 30 and 3000 LY in dimension. It has the merit of accounting for the existence of stable clouds of such a mass that self-gravitational forces alone would be insufficient to bind them. Indeed, this magnetic instability can be about 100 times stronger than just self-gravity.

Now that we have our Galactic-scale spiral structures, threading massive "supercomplexes" of clouds, in which we can generate individual cloud complexes, we can consider the life of a nursery-sized cloud.

7.4. Forces on a cloud

To describe the physics of a stellar nursery we need to know the extent, total mass and temperature of the cloud; the degree of ionisation (in other words, how many of its atoms have their full complement of electrons, rendering them neutral and how many have lost one or more electrons, leaving positively charged ions and negatively charged electrons upon which the

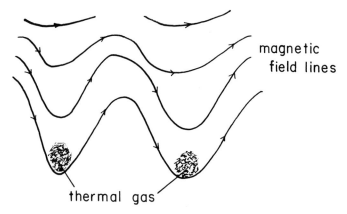

Figure 7.1   The Galactic magnetic instability that leads to the clumping of gas from which dark cloud complexes form.

magnetic field can exert an influence); the amount of rotation in the cloud; and the strength and geometry of whatever magnetic fields thread the cloud. It would also be valuable to know the relative proportions of atomic gas, molecular gas and dust. Then we can investigate the gravitational effects as one piece of the cloud pulls on another; any electrical forces present between ions and dust particles that could lead to relative motions of gas and dust; the degree of influence exerted by magnetic fields on gas and dust particles (none unless some particles are charged or ions are present); the effect of spin in dictating the size and shape of the cloud. As you can tell from even this rather short list, clouds are complicated structures. It is only with the advent of larger and larger computers that it has become feasible to treat the collapse and evolution of clouds in any kind of realistic physical detail.

Let us consider the cloud cores for a moment. What relationship do they bear to the original, extensive dark clouds? How did they become "cores" and are they still connected in some way to the rest of their parent clouds? After the self-gravity (the mutual, relentless pulling of one piece of a cloud on all others) of a cloud has done its work, the cloud is somewhat clumped in its centre. Radio molecular data show that these centres are dense and that, as we move away from them, the density falls somewhat like an inverse square law (i.e. you get 1% of the initial density at 10 times the initial distance from the geometric centre of a cloud). The rest of the cloud is in an envelope; a halo around the core. It is interesting to note, however, that the average cloud core doesn't seem to be collapsing gravitationally. This may seem odd at first – after all, how else are we to manufacture new stars? But if cloud complexes everywhere were all to collapse bodily, the resultant Galaxywide rate of star formation would far exceed the measured rate. In other words, even once cores have formed in a cloud, their further fragmentation to make stars takes a long while, and the entire procedure, from cloud to star, probably proceeds through a number of phases in each of which it will appear as though little is happening. These are termed "quasi-static phases", for obvious reasons. It is both instructive and relevant to remember that, although we have noted several cases of outflows, on different spatial scales, among very early phases of low-mass stellar evolution (Chapter 6), there's not a single known situation of unambiguous inflow of a cloud complex, or even a chunk of one! From this we might hazard the guess that either, when infall of large chunks of clouds occurs, as it must on some scale, it lasts for a relatively short time, thereby reducing our likelihood of catching it in action; or we don't need a concerted rapid inflow of an entire cloud complex, or even of a sizeable fragment of one, but only of the smallest relevant subunit of a cloud – namely a core.

What kinds of force, then, are applicable to cloud cores? As radio instruments either become larger, or arrays come on the air at the crucial millimetre wavelengths, where cloud molecules are observable, we find that clouds are actually not smooth entities but are resolved into lots of smaller pieces. These pieces are often comparable with the size of the entire Taurus-Auriga nursery – almost ten thousand solar masses. Each nursery-sized clump must be supported against its own collapse or else radio astronomers would already have detected inflows of some clumps. Internally, already-formed young stars stir up the clumps, generating broad scale turbulence. This chaotic running around and colliding of bipolar (and less well-organised) flows creates an important internal pressure (motion is like temperature which provides thermal support) that can inhibit clump collapse.

Magnetic fields, too, can play a critical role. These can force any internal motions that might foster collapse to take place only along the field. In other words, given only a relatively tiny ionised (or charged) component to the largely neutral cloud material, the field prevents ions from moving across magnetic field lines. Another handy task is "magnetic braking" of clouds, whereby fields can very efficiently dissipate the angular momentum of an originally rotating cloud. This is achieved through the intermediary of a magnetic wind from the star, in which the magnetic lines of force tie the fast-spinning protostar to its much more slowly rotating cloud envelope, exerting a substantial slowing torque. It is, therefore, interesting to note that our favourite complex, Taurus-Auriga, does not evince any signs of large-scale rotation and does reveal evidence that magnetic fields have influenced the directions of collapse of subunits of the clouds. Clearly, the fields do not inhibit the formation of low-mass stars. We might, however, be tempted to muse on the absence of high-mass stars in Taurus-Auriga and perhaps even to speculate that fields could have prevented their formation (by hindering the collapse of more substantial clumps of material than it takes to create a T Tauri star).

What about rotation itself? In principle, a cloud can support itself against gravitational collapse simply by rotating. You can naively imagine how difficult it would be for a stone, whirling on a string around your head, to fall in and hit you. This centrifugal support is available to a cloud core too. Unfortunately for the proponents of this idea, cloud envelopes are observed to rotate only rather slowly and cores present too small an area to have been studied yet in this way. However, if cores are subject to the quasi-static mechanism that we shall describe in a moment, we can expect the magnetic braking, mentioned above, to have kept down their rotation so that cloud and envelope spin together, as a single unit. Turbulence, magnetic fields, rotation – the "big three"; among these processes lurks the answer to the question of the support of cloud cores. Current thinking has turned increasingly to magnetic methods for core support so let's examine the idea in a little more detail.

7.5. The loss of magnetic fields

Magnetic fields control the motions of ions. In dark clouds, not studded with hot, young, high-mass stars, there is no major source of ultraviolet photons to provide ionisation. Therefore, most of the mass in a cloud remains electrically neutral and, as such, is not directly affected by magnetic fields. But the stream of ions and electrons that are controlled by the field suffers collisions with the bulky neutral particles ("neutrals"). These collisions serve as a frictional coupling, rather like a clutch, between ions and neutrals. In short, the field indirectly can steer even the neutral cloud material. So the neutrals are not entirely free to go about their business. The neutrals feel the self-gravity of the cloud – they, at least, are trying to collapse. But, through this coupling, the field provides partial support of the neutral body of a cloud, hindering any attempted collapse. Fine, you may say, then how do cloud cores ever collapse? The coupling of field and neutral material depends for its existence on a slippage between the ions (to which the field is tied) and the neutrals: no slip, no friction – no coupling. Therefore, by providing this partial support of the cloud, the field also promises to slip away, out of the region of increased density in the core, and into the surrounding cloud envelope. It is a long, slow process but, as the field gradually leaks away, the neutrals are left behind, under self-gravity, and they congregate into even denser cores than before. Once this has happened, there is very little else that can bolster up the cores against gravity's relentless tug. Turbulent pressure also diminishes with the loss of the field; only thermal support is left and gravity never relaxes its grip. The production of dense cloud cores is, therefore, inevitable.

Recent observations have provided definite evidence for "quiet cores"; that is, density enhancements, in the same clouds where T Tauri stars are seen, that are not very turbulent. The physical characteristics of these small cores (their densities, sizes, temperatures) indicate, most excitingly, that these entities are on the very brink of gravitational collapse! Some of these cores even contain embedded infrared objects – protostars, perhaps? The overall picture is taking real shape now. Incidentally, you can find from this theoretical approach that there is no physical difference, no obvious break, between cores and their envelopes; one smoothly blends into the other.

Magnetic fields have done their best to support the cloud cores against gravity but, eventually, this is no longer sufficient. The diffusion of fields away from dense centres guarantees the production of very fertile quiet cores in which collapse, and hopefully star formation, can occur. This process of leaking fields leads to dense cores on a timescale of one to ten million years. The ages of T Tauri stars (Chapter 7.2) are of the same order, which may suggest that the fertile cores are as abundant as these stars in stellar nurseries. If so, the construction of radio telescope arrays, designed to operate at millimetre wavelengths, should yield copious examples of these missing links between clouds and low-mass protostars. A timescale of millions of years might also bear upon the very broad spread found in the ages of T Tauri stars, even within a single nursery (Chapter 7.2). Finally we have to wait a few hundred thousand years, once cores start to collapse, before the production of the earliest visible new stars.

7.6. Protostellar accretion: what stops it?

Since we have stressed the point that cores have not in any sense "snapped off" from their surrounding molecular envelopes, how does a collapsing cloud know when to stop? Why doesn't a cloud try to make a 100 solar mass star? How does it know that 1 solar mass is the "right" kind of size? How is the relatively low efficiency of the star-forming process incorporated into the scheme?

Chapter 6 showed us that the youngest low-mass protostars of which we are aware already are busily engaged in flinging material back into the clouds. Likewise, molecular observations of somewhat higher-mass infant stars reveal bipolar flows out into the cloud envelopes, although without the elegantly confined and collimated structures of which the pre-T Tauri objects are capable. It appears that the infall has been terminated by switching on rather powerful stellar winds even in these very young objects. Then the basic issue becomes that of determining the powerhouse for these winds, for it must be the winds that dictate when "enough" mass has accumulated in the protostellar cores.

What do the observations tell us? They reveal that when we are first able to study low-mass young stars visually, that is once they have become recognisable as T Tauri stars, they are already fully convective objects (Chapters 2.4 and 5.4). Their early visibility speaks for the vigour with which the winds have swept clean the obscuring circumstellar dust. Their convective nature suggests a potential process responsible for the onset of winds. Let's clarify the importance of this convection. Remember that this tells us that any gravitational energy gained by collapse of the protostellar core is communicated to the stellar surface by seething, boiling, bubbling motions ("convection": just as water, boiling in a pan, communicates the heat from the bottom of the pan to the surface). Further, these convective motions act to produce a dynamo by the interaction of their twisting internal motions and the pre-existing magnetic field that has been "torqued up" during the formation of the protostellar core, following the conservation of angular momentum in the collapsing core. What amplifies the strength of the dynamo is the "differential rotation" of the protostar. This implies that different layers and latitudes in the stellar configuration rotate at quite different speeds. Even today's sun does this, and Jupiter too: their equators rotate much more rapidly than their polar caps. Now the cloud core that dumped its material into this fledgling star knew nothing about what it was constructing; in particular, it could not have realised that, for a star, the minimum energy configuration is one of uniform rotation (everything going as closely as possible at the same speed). So the cloud created a differentially-rotating object, but one that, initially, actually was radiative and stable against convection. But a short time later we are to see T Tauri stars that are convective. What can happen to alter so radically the means of communication of energy between core and stellar surface?

7.7. Chronology of accretion

To tie all these diverse ideas together, let's try to follow the chronology of a cloud core that is collapsing to a star. Not all the gravitational energy given up for the core to collapse is radiated away to, and by, the protostellar surface layers. An appreciable quantity of this energy can be hidden in the form of differential rotation. Material of low angular momentum collects at the centre; material with high angular momentum cannot give up its spin and must, therefore, collect into a rotating disk that surrounds the fledgling star. Still the matter rains down onto the forming star creating, not surprisingly, a shock: a shock that arises because of the constant collision of the growing stellar surface layer and the still infalling, accreting matter. This shock radiates a substantial fraction of the luminosity of the early protostar. The accreting matter is dusty and absorbs optical and ultraviolet radiation, hiding the star, and re-emitting this energy in the infrared. We have manufactured a purely infrared object. At some distance from the star, however, the dust will be heated so greatly that it will evaporate, leaving a region down to the stellar surface that is clear of dust. However, from the outside, we see only the low-temperature "dust photosphere" and cannot probe beneath it. This dusty exterior is about 1500 solar radii from the star's centre, while the dust-free zone is about one tenth of this in radius. To put it in its place, the accretion shock is located only about 4 solar radii from the centre of our newly-forming star. Theoretical models have been constructed of these protostars and one rather unusual aspect of this problem is that these objects' outer layers are heated from above by the radiation from the shock! (Conventional stars have the decency to be heated only from their insides!)

Now, in the core, an event takes place that changes this situation and is capable of providing appreciable heating from within, too. As we discussed in Chapter 2, hydrogen fusion into helium necessitates a temperature of some 15 million degrees for its inception. Our low-mass protosuns are not going to achieve that for a long time. However, they do manage to attain about one million K and that is sufficient to burn deuterium. This is part of the process by which hydrogen burns to helium: two nucleons (one proton, one neutron) constitute the "deuteron" or the nucleus of "deuterium" or "heavy hydrogen". When we discussed hydrogen fusion in Chapter 2.5 and 2.6, we treated the fusion of the four protons into helium as a single process. However, it really proceeds by stages. First, two protons are converted to one deuteron, the nucleus of one of the "isotopes" of hydrogen. An isotope is another version of an atomic element with a closely similar number of nucleons. Hydrogen has two isotopes: deuterium and "tritium", the latter consisting of three nucleons. Helium, too, comes in two isotopic forms: the common helium we've already encountered with four nucleons, and a three-nucleon version too. So, our deuteron is then expected to combine with a passing proton to be converted into "helium three", the three-nucleon isotope. Only if this helium-three nucleus meets another of its kind fast enough to overcome the substantial electrostatic repulsion between the two identically charged nuclei can we achieve the final stage in making helium-four. It is this final step that requires the 10-15 million degrees that we spoke of in Chapter 2. But, at a temperature of only 1 million K, we can still go as far as "deuterium burning", yielding helium-three, but no further. Nevertheless, each of these reactions along the way, that fuses protons successively (in what's termed the "proton-proton chain"), liberates binding energy equivalent to lost mass, exactly as we saw for the fusion of helium-four. In fact, deuterium burning develops 30% of the total energy of fusion to helium-four, not too shabby an amount when it occurs inside a developing protostar. In response to the onset of this central heating process, almost the entire star becomes convective and begins to seethe, boil, and turn over internally.

The convection proceeds in an unstable manner – the more the core heats, the more it drives the convection, the more it tries to drive further convection. Now we have our extra ingredient: a differentially-rotating, convectively-unstable star creates an immense dynamo, as magnetic fields are dragged and twisted up to the surface where they create havoc for the star, and all the wonderful phenomena that observers of T Tauri stars delight in watching (like flares, winds, spots)! This situation persists until all the extra mechanical energy (i.e. the difference between what is stored in the original differentially-rotating state and the minimal energy that could be stored in the uniformly-rotating state) is released.

After all this turmoil, we have a convective star; one that is optically visible, has attained a rotationally less energetic internal state, and has halted its accretion. It stopped at whatever mass it needed to accumulate in order to raise its core temperature high enough to ignite deuterium burning. For the existing calculations of an ultimately 1 solar mass star, once the core is 0.4 of a solar mass, it is at about a million degrees, and once it has increased to 0.5 of a solar mass, it's virtually entirely convective – an almost fully-fledged T Tauri star. It appears on the "birthline", at the top of its convective track in the HR diagram, because it has generated a wind that halted accretion at 1 solar mass, and has dispersed its previous dusty circumstellar cocoon. The age of this young sun would be about 100,000 to 1 million years from the initiation of cloud core collapse. We should note here that these accretion times were not included in the interpretative HR diagram that was used to derive ages for T Tauri stars. Strictly, we should correct the ages of the youngest of these stars by perhaps 100,000 years.

Our journey from cloud core to T Tauri star is complete!

8

Towards a more sedate life

And quiet did quiet remain.
W. De La Mare
The Song of Finis

Here, where the world is quiet;
A. C. Swinburne
The Garden of Proserpine

Beyond T Tauri stars – stellar ages and lithium – hydrogen and calcium emission lines – rotation

---

8.1. Beyond the T Tauri phase

When we left the T Tauri stars, at the end of Chapter 5, they were busily engaging the attention of astronomers who work across the entire electromagnetic spectrum. When do these vigorous stars see the error of their youthful ways and grow up? How do they approach the main sequence where (almost) all is quiet and relatively placid? Only very minor flares and chromospheric activity attend these mature stars. What are good quantitative criteria by which to assess the evolving activity among T Tauri stars and their successors?

The powerful "hydrogen-alpha" emission lines can be expected to linger for a while as T Tauri stars moderate their behaviour. Likewise, and perhaps even more useful, since even our sun shows local enhancements of calcium, there are blue lines of ionised calcium that persist in emission in even the oldest T Tauri stars (20-30 million years). How do these spectroscopic symptoms of atmospheric activity die away?

8.2. The problem of stellar ages

So far, when we have discussed the youngest (low-mass) stars, we have used ages derived from the interpretative HR diagram (Fig. 5.9) and we have had to modify only the very youngest of these (much less than a million years) to allow for the comparable time of the accretion phase (Chapter 7.7). Now we find ourselves moving to the other extreme, where we need to be able to distinguish stars that are a hundred million from ones that are a few thousand million years old. How do we date an individual, considerably post-T Tauri, star?

We mentioned in Chapter 5.1 the significance of strong lithium absorption lines in the spectra of T Tauri stars: lithium is an element that does not last for very long inside stars. Therefore, its strength in T Tauri stars indicates their youth. It is also seen in much older stars, although it is, of course, far weaker than in the T Tauris. In principle, one can assess the amount of lithium on the surface of any star, decide upon the rapidity with which this is destroyed in a star of the appropriate temperature, and estimate, from an assumed initial amount of the element, for how long lithium has been undergoing destruction. This gives us a "lithium depletion age", subject, as you can probably tell, to several uncertainties. However, it may be all that we have for random stars. To compare very young and much older stars we should really increase the ages of the main sequence stars by a period of time to represent their contraction times too, say 50 million years. This unites our two methods of dating young and old stars.

Then we are ready to investigate the atmospheric emission-line activity in both T Tauri and in main sequence stars. Of course, we do have one major problem with our stellar sample for study. Recall (Chapter 5.7) the surprising paucity of likely "post-T Tauri stars". Our analysis of the evolution of stellar behaviour may have to depend upon the comparison of two radically different classes of object with little, or nothing, to represent the epochs between the infant T Tauri stars and the mature main sequence objects.

8.3. Hydrogen and calcium lines

To isolate effects dependent upon stellar mass, we should compare roughly 1 solar mass T Tauri stars with stars of the same mass on the main sequence; namely, solar-like stars (for the aging of T Tauri stars is essentially along a horizontal path in the HR diagram, when their temperatures grow from about 4000 K to about 6000 K, with relatively little alteration in luminosity). It is true that the surface temperature of a young star grows appreciably along its radiative track. However, it is believed that the agency responsible for hydrogen and calcium emission in T Tauri stars does not depend on photospheric radiation but rather reflects subsurface magnetic and convective phenomena. So investigators have chosen to look at the luminosities of these emission lines to compare different stars with one another.

The first gratifying thing to note is that hydrogen and calcium are well-correlated in stars which show both. That suggests that even when one of our indicators of activity fades away (hydrogen goes first), the one that is more persistent (calcium) will still guide us accurately.

For older stars, up to 2 thousand million years in age, there is a clear decline in hydrogen strength with age. On the main sequence itself (we can, of course, use our own sun's measured activity here to represent a fiducial point with known age, 5000 million years), hydrogen-alpha activity decays with time again, apparently halving in about 4000 million years. Can we connect the T Tauris' behaviour to the very limited activity of their much older counterparts and attribute the dramatic difference solely to evolution? Our T Tauri stars sport hydrogen lines up to 100 times greater than even the most active main sequence objects! It is, therefore, possible that the hydrogen-emission activity in the strong-line stars does not come from merely a more powerful version of the mechanism that operates in main sequence stars. Given the lack of post-T Tauri objects, it is tantalising to speculate that the T Tauris very rapidly drop in surface activity, perhaps to a background level more in keeping with the main sequence stars.

It has been argued, from young (though not T Tauri) stars in a somewhat older cluster than our nursery in Taurus, that calcium emission weakens as stars age in accordance with two different time scales. Perhaps there are indeed two quite different phenomena that characterise the behaviour of young and of moderately old objects. This calcium study indicates that one effect halves itself in about 60 million years: this would be the rapid effect. The second falls much more slowly, halving its activity in about 8000 million years: that would be the relevant phenomenon for main sequence stars.

8.4. Stellar rotation

We ventured into issues of stellar rotation among the T Tauri stars in Chapter 5.6; we will discuss these rotation speeds again in Chapter 10.4 because they will bear upon the possibility of forming planetary systems. For the moment, our concern is with any systematic changes in rotation, as we leave the youthful T Tauris and head for the main sequence. There is an interesting pattern along the main sequence whereby one can divide stars into fast and slow rotators, with the division occurring for stars a little more massive and warmer than our sun. Above this mass, stars are all rapid rotators; below it they are slow rotators. We shall speculate in Chapter 10 on a possible explanation for this in terms of the length of a star's convective track during its pre-main-sequence phase. Aside from this break, for solar mass stars it has been suggested that the rotation depends on the inverse square root of the age. So given two stars of ages 1000 million and 100 million years, the older would now be rotating about 3 times more slowly than the younger. Indeed, this relation works quite nicely for stars all the way from about seventy million years of age to a few thousand million.

However, rather recently, a bizarre phenomenon has been noted in the youngish cluster, the Pleiades (alias "The Seven Sisters"): a pretty, compact group in Taurus which is about 70 million years old. Most of the stars in the Pleiades are not rotating perceptibly, in agreement with the square-root-law mentioned above. But as many as 30% of the sample studied are spinning faster than predicted, some as much as 10 times faster! These stars are heavily spotted. It is argued that they have only rather recently arrived at the main sequence, and have actually been spinning up (increasing their rotation speed) whilst pursuing their radiative tracks! This curious effect is claimed to be accounted for by changes in the internal mass distribution of the stars. For stars in the relevant mass range, the radius actually decreases by about one third along the radiative track. In addition, the radiative core becomes more compact. These combined changes are estimated to alter the mass distribution in such a manner that conservation of angular momentum (Chapter 5.6) would demand that the equatorial speed of rotation should increase by a factor of 3 (assuming that the entire star rotates as a solid object and barring any substantial braking by our old friend, the magnetic field). Such an increase could account for many (but not all and, in particular, not the fastest) of the anomalous fast rotators in the Pleiades. It is also noted that some of the slightly more massive stars in the cluster have had more time to have undergone magnetic braking, and indeed these are not rapid rotators, in spite of any spinning up they have undergone.

In short, we find a rather erratic history describes stellar spin. During the earliest years, magnetic braking de-spins rotating cloud cores. Then T Tauri stars spin down, rather than up, along the convective tracks for reasons that we have not yet identified (but will consider in Chapter 10.4). Subsequently, stars on their radiative tracks spin up (perhaps with hindsight, we can now understand this effect), only to spin down again, magnetically braked, after they first arrive on the main sequence. Thereafter, stars behave themselves, and spin more slowly. (At the time of its formation, our sun could have revolved once on its axis in only 3 hours; now it requires 28 days!)

8.5. T Tauris again: both ends against the middle

There is a philosophical issue that confronts us when we try to understand the T Tauris and, particularly, the pre-T Tauri objects. Should we look at the zaniest HH-exciting stars and try to elicit, from their extreme behaviour, the characteristics of "normal", or at least more typical, T Tauri stars? Or is it preferable to examine the quieter members of the class in which the less well-understood earlier phenomena have long since died away? An identical dilemma faces us when trying to extrapolate from the

T Tauris to the main sequence stars: do we work forwards or backwards? Each approach has its merits. My personal opinion is that two very different types of behaviour are involved in any T Tauri star: the more extreme dominates in the youngest stars; the less drastic in the "older" objects. Similarly there is a striking parallel in the suggestion (Chapter 8.3) that phenomena on fast and slow time scales characterise surface activity in stars: the more rapid presumably responsible for the hasty conversion of the T Tauris into mere shadows of their former selves; the slower operating primarily on the main sequence.

Perhaps much can be made of the depth of the convection zone in different evolutionary states. During late protostellar and early pre-main-sequence evolution almost all of a stellar interior partakes of convective motions. These internal bubblings subside greatly as stars age along the radiative track, and the convection zone in the modern sun is only skin deep. During these epochs, convection does much to keep the interior and the atmosphere of a star in touch with each other. It is, therefore, probable that convection, with its attendant wrapping, twisting and rising of stellar magnetic fields and associated stellar winds, is behind the steady maturation of stars and the "settling down" of their outer layers.

This is as far as we shall go in our examination of stellar evolution. We have watched our sun from the epoch long before it was even a protostar, through its chaotic earliest 500,000 years, past the 20 million years or so of life as a T Tauri star, and have left it where it was, several thousand million years ago, having achieved its haven on the main sequence. Now we shall go back in time, again, to do justice to the high-mass stars and to their particular methods of formation (Chapter 9), and to pursue the issue of how our planetary system arose (Chapter 10).

9

High-mass stars and triggering mechanisms

What can we reason but from what we know?
Alexander Pope
An Essay on Man

High-mass stars – the birth of O-stars – sequential star formation – the supernova trigger – the first stars

9.1. "How the other half lives"

It is time to look a little more closely at the birth of high-mass stars. Clearly their story cannot bear directly upon the birth of the sun, or of the earth, but some stellar nurseries apparently manufacture principally high-mass objects, which is curious, if nothing else. Too, in these O-associations there could be T Tauri stars and it would be interesting to see whether the conditions of their formation differ from those in the less flashy nurseries that we have already considered.

There are global issues involved, too, for O-associations are confined to the very largest molecular cloud complexes that are strung along the spiral arms. The compression of the interstellar medium, following shocks associated with the density waves, would then be responsible for creating these luminous blue stars, according to some astronomers. However, recent molecular studies of the giant complexes have led others to suggest that these great clumps of gas and dust are so thick and so large that they are quite unlike the very tenuous, fragile medium that was once envisaged as being easily compressed into forming stars by modest shocks. Perhaps the interiors of giant molecular complexes are largely buffered against any of the external pressures that have been proposed as providing the "last straw", that could push a cloud over the brink or trigger it into collapse when it was previously only considering collapse. In this scenario, star formation interior to the complexes must occur fairly steadily, unperturbed by any external circumstances. Yet the high-mass stars must still be confined to the spiral arms, though low-mass objects could form in clouds not tied closely to the arms.

9.2.   The birth of an O-star

There is little disputing the tremendously disruptive potential of already-formed O-stars on their parent clouds. When one of these luminous, hot objects appears in a molecular or atomic gas cloud, it wreaks havoc on the cloud. The dense hydrogen close to the star will become ionised, forming a very compact HII region (Chapter 3.2: usually observable in the radio region), where the density and temperature will both be enhanced over the ambient medium. This physical contrast drives the ionised gas out into the cloud by virtue of its greatly increased thermal pressure. Again, the pushing of the cloud gas is at around 10 km/sec, rather like the CO bipolar flow velocities near pre-T Tauri stars (Chapter 6).

9.3.   Sequential star formation

What might this ionisation front achieve? Were the rest of the cloud to contain sizeable subchunks that might like to collapse one day, this expanding pressure wave might provide a final shove, and there is definite evidence that these complexes are truly quite lumpy. Thus it would, in principle, be possible to initiate a self-sustaining wave of O-star formation throughout the entire cloud complex. High-mass stars are believed to form in clusters, rather than in isolation. Sometimes it is even possible to observe a sequence of groups of O-stars that have formed, one after another, near the boundaries of a molecular complex. This sequential formation of high-mass stars has been proposed to account for many of the recently-formed and forming O-stars. It has been likened to an infectious disease, gnawing away at cloud complexes and eventually overrunning them.

Watching such a process at work on a cloud, we'd see several simultaneous stages in the development of stars. First, we'd find full-grown young clusters of massive stars at the outside of the cloud, then a region between them and the cloud where a wave of ionisation is advancing into it, driven by the ultraviolet radiation of the already-formed stars. Finally, we'd see the dark cloud itself, rich in molecules, readily detectable by microwave radio emission, and studded with sources of infrared radiation. These infrared sources represent the birth sites of new high-mass stars which are still heavily shrouded by dust. This process operates sequentially and is self-perpetuating, given an initial star of high mass, since it preferentially produces high-mass stars from the cloud.

Recently, very detailed computer simulations have been performed to study the effects of the waves of ionising radiation from O-stars on clumps embedded in the surrounding molecular material. These new calculations draw heavily on the observations of quite small-scale inhomogeneities in dark clouds, establishing clumpiness on scales less than 1 LY across. The radiation of a single O-star has been shown to be effective at "imploding" these clumps, although no globule of material was created that was massive enough to collapse gravitationally, by itself, and to form new stars. Since O-stars are so often found in clusters of their fellows, perhaps a clump might be squeezed simultaneously from two sides by the radiation of two distinct hot stars?

These fascinating new calculations show the way in which the double blast of radiation first ionises, then evaporates away, much of a compressed clump of gas. If the original stars were too hot, the entire clump could be lost to evaporation. But, for slightly cooler O-stars, the compressed clump of gas can survive long enough to collapse onto itself if self-gravitational instability sets in (this instability depends on having more than a critical amount of mass present – the "Jeans' Mass", after Sir James Jeans, who investigated the equilibrium of massive gas clouds). Within a few tens of thousands of years, then, several pre-existing O-stars could squash gas clumps into collapse to form a new generation of their kind. These new stars would implode other blobs of cloud material in the same manner, leading to an entire cluster of newly-formed O-stars within a region only a few light years across. Thanks to the VLA (Chapter 6.7), there now exist radio images of small clusters of the very compact ionised regions that surround newly formed

O-stars. These maps suggest that stars have formed within only 10,000 years of each other in entire groups of massive objects. Because of the evaporation of small globules, the multiply-squashed clump idea leads only to the formation of new high-mass stars and does not compress chunks to make low-mass stars. It is possible that, at great distance from several O-stars, smaller masses of gas could be encouraged to collapse, but the dominant effect is still the self-propagation of massive stars. Both the newest calculations and observations actually run counter to the predictions of the original "sequential star formation" concept (where a single advancing ionisation and shock front was thought to induce collapse in a much more uniformly dense cloud). So, in order to be fashionable, perhaps we should switch our bets and back instead the "multiple radiation-driven implosion" theory!

9.4. The supernova trigger

Another idea for "triggered" star formation is via the supernova trigger. Proponents claim it may be capable of yielding stars over a wide range of mass – which more accurately fits the cosmos as we see it than the production solely of high-mass stars. The mechanism goes into action when the high-speed shock wave and debris from a dying star – a supernova – expand out into the interstellar medium. Before its momentum is lost, the shock wave will sweep up a large volume of slowly-moving material (conservation of momentum). Proponents of the model argue that eventually the blast may compress the mixture of pre-existing clouds and ploughed-up matter to such a density that self-gravitation can take over and finish the process of star formation.

The antecedents of the idea go back about thirty years to the prophetically perceptive astronomer Ernst Opik. Years later, Rudolph Minkowski wondered about the intriguing arcs and chains of stars (newly-born?) seen at the rim of the Veil Nebula in Cygnus, a giant gas-shell remnant of a supernova explosion. However, what really stimulated a resurgence of the supernova trigger model were studies of a large expanding shell of ionised gas in the constellation of Canis Major. At one edge of this shell lies a complex of dense dark clouds, coupled with stars, some of which still illuminate the material from which they presumably formed. It was concluded that a supernova had triggered the clouds at the shell's edge into collapse and produced the associated young stars.

There is still controversy about this model and several unanswered questions linger. While the Canis Major shell may be the product of a recent supernova, it does not have the radio emission spectrum usually associated with one. It might have lost this in gradually merging with the interstellar medium, but in that case there is no clear evidence for any amount of neutral gas that ought to have been swept up. Nor has any pulsar, the spinning collapsed remnant of the star that blew up, been found within the region – nor any other identifiable stellar remnant.

The stars in the clouds on the edge of the ionised shell of gas do not appear any younger than other "field" stars found in the general region, which were presumably not triggered into collapse by any external force. Even if the stars at the edge of the gas shell were young, this ignores the delay that might occur between the arrival of a blast wave and the initiation of collapse, essentially the time for a freely-falling mass to reach the cloud centre. If a blast wave triggered the birth of stars, then these should form some distance away from the supernova itself and, the older the expanding gas shell, the farther from the centre should be the stars created by the blast. But this displacement was not found either in the Canis Major region or in other regions where supernova-triggered collapse has been suspected.

Despite these questions there is a curious aspect to star formation even in the Taurus-Auriga nursery that bears upon the possible existence of an external timer. The structure of these nearby clouds (Fig. 3.8) resembles a series of beads on a wire, the "wire" being a broad elongated ridge of gas and dust and the "beads" denser blobs in which young stars are visible. Study reveals that the stars in each bead appear to be about the same age. If the supernova trigger model is correct, how could one supernova explosion have controlled the formation of stars along this ridge – almost 60 light years in length? Neutral hydrogen maps show an expanding shell of gas with one edge approaching us and the far side expanding more slowly into the Taurus clouds. Its shape and internal motions could suggest a supernova explosion, and the deceleration of material on the rear part of the shell may indicate that it is interacting with the densely-crowded stellar nursery.

In a further investigation of the supernova trigger idea, a large nebulous loop in the constellation of Monoceros was selected as a testing ground. It is a convincing supernova remnant from the shape and radio emission of the shell. It was then necessary to establish the physical proximity of the dark clouds and the expanding gas shell. As another prerequisite, interaction between clouds and gas shell had to be clear-cut. Furthermore, any delays between the arrival of the blast wave from the supernova and the epoch when the first stars formed had to be investigated. Lastly, it had to be shown that stars in those clouds which had suffered disturbance were young enough to be a by-product of the blast.

Photographs of the Monoceros Loop taken in the light of ionised hydrogen clearly reveal small, dense, dark clouds throughout – some apparently close to the luminous remnant's rim and others projected against its inner portions.

Firm evidence was found that the gas in the globules inside the loop has been disturbed by comparison with the gas in the globules which have yet to feel the blast wave.

No young stars have been found yet, in the disturbed cloudlets, at optical wavelengths which would enable an age to be determined for them. Perhaps not enough time has passed for the creation of fully-fledged stars, and this study still continues.

9.5. The final word?

Our detailed and rather complex description of the events that take us from a cloud core to a protostar (Chapter 7) really do not address high-mass star formation. By their very nature, high-mass objects are quite differently behaved than low-mass stars. Their ultraviolet ionising flux is prodigious; their mass flows and the velocities of their winds are far greater than those of even pre-T Tauri stars. It is, therefore, plausible that the generation of these massive stars could only be mediated by processes of equal violence such as the globule-compression model.

Using the unusual circumstances of low-mass protostellar behaviour as a yardstick, it is surely significant that HH objects are not found in association with forming O-stars; nor are spatially resolved, thin radio jets; nor spatially well-separated narrow CO caps; nor dust zones as flat, and as small in radius, as those around HH-exciting stars. It is, in my opinion at least, very probable that quite different circumstances attend the births of high- and low-mass stars. Certainly the elegant collimation of stellar mass loss shown by the comparatively very low luminosity pre-T Tauri stars is not achieved by O-stars, which seem to incline very much more to a "brute force" approach to the dispersal of their dusty and gaseous prestellar cocoons.

9.6. The first word

When the Universe was young and galaxies too were young, the medium that pervaded those galaxies was essentially all hydrogen and helium. The heavier elements were yet to be cooked up in stellar cores by nucleosynthesis (fusion reactions), but there were no stars at that early epoch. How did the first stars form and what were they like? One can only surmise that the first were stars of almost pure hydrogen; were of great mass, probably much heavier than the 100 solar mass limit that we would impose nowadays, for no dust grains existed to allow cloud fragments to contract to small sizes to make solar-type stars. Hence, there must have been a special population of very large stars whose interiors furnished the heavy elements and dust grains that permitted the next stellar generation to be born with radically lower masses. Of course, this very early galactic population has entirely vanished from the modern Universe, but we owe it a great deal (no less than our very existence!).

We now return to the confines of our own solar system and investigate the prospects for the manufacture of extrasolar planets.

10

Planets

The disks revolve, they ask to be heard –
Sylvia Plath
The Courage of Shutting-Up
(winter trees)

Observe how system into system runs,
What other planets circle other suns.
Alexander Pope
An Essay on Criticism

Circumstellar dust disks – protosolar nebulæ – timescale for planet formation – HL Tau – big disks around HH-exciting stars – observability of planets – rings and moons

10.1. Introduction

The detection of planets outside the solar system is indeed a difficult problem, even by indirect methods. It is crucial to envisage how planets arise as a natural consequence of star formation in order to understand the specific details by which our own solar system may have formed. For astronomers whose interest focuses upon even the closest nurseries of stars, the first issue is not the recognition of an individual planetary object, but rather the convincing demonstration of the existence of a flattened, disklike distribution of material surrounding a young star. From this material one can readily imagine that planets might one day be born. So there are two problems: first, to find evidence for "circumstellar" dust; second, to learn the geometry of this dust. Clearly, all that we have so far seen about the T Tauri stars leads us to recognise that these stars do have circumstellar dust shells. It is unravelling the morphology of these shells that is the task at hand.

10.2. Circumstellar disks

We saw at the end of Chapter 7 that the collapse of a chunk of cloud into a young star is a rather messy process. Of necessity, the fledgling star has a lot of loose ends to tidy up. Much material has to "rain in" on this incipient star before we can see the starlight escaping from its confining dusty cocoon. Rapidly rotating matter from the original cloud cannot fall directly onto the star since it must preserve its spin. It must create a spinning disk around the star in which material queues up, in spiralling orbits, to be swallowed by the stellar core. Now what happens to this disk?

The disk serves as a repository for all the collapsing material that has too much angular momentum to fall directly onto the protostellar core. As we saw, the very vigorous young stars, of mass comparable to the sun's, are attended in their youth by quite large dusty disks. When we see what we believe are later states of these same stars, as visible T Tauri stars, there does seem to be evidence suggestive of dust disks. But these are now mere vestigial remnants of the substantial flattened systems that surrounded the predecessors of the T Tauri objects, namely the exciting stars of HH nebulæ. What other changes accompany the seeming disappearance of the early disks?

In Chapter 5 we discussed stellar rotation among the T Tauri stars. As these stars evolve along their convective tracks they appear to spin down, rather than speed up. Something is carrying off stellar angular momentum. As with ice skaters, these stars must "extend their arms" to rotate more slowly; that is, some of their material must be flung out to great distances or some of their distant, pre-existing circumstellar matter must be encouraged to rotate more rapidly. The orbits of some disk particles bring them into the star where they are accreted, while others spiral outwards, shed by the protosun. We have an active transport of angular momentum within the disk, both inwards and outwards. If we seek guidance from the present-day solar system, we find that 98% of the system's angular momentum now resides within the orbital motions of the planets, leaving only 2% in the form of solar rotation. This explains why it now takes the sun about one month to rotate once on its axis (it actually takes 34 days at the pole, but only 25 days at the equator as a consequence of its differential rotation) whereas, theoretically, the protosun could have spun once in only a few hours! Do protostars, therefore, shed their unwanted angular momentum into their circumstellar disks? If so, how is the momentum fixed? It's no use invoking hordes of small dust particles, or rings or blobs of gas, for all these carriers are subject to redistribution or disruption by other processes. However, one form of redistribution turns out to be positively constructive. Some small particles collide, slowly, with their neighbours in the disk, and these can stick to one another. This kind of mechanism leads to the accumulation of larger bodies. These larger bodies begin to influence the small grains and gas that lie in the disk in their vicinities. Steadily these protoplanetesimals stake their claims to finely divided material at some small distance both interior and exterior to their orbital radii. A ring system like Saturn's is a perfect example of this kind of process, where material spirals both into the planet's globe and away from it, and where the gravitational influence of moons and moonlets can be recognised by enhancements and rarefactions in the surface density of the disk. These contrasting regions of the disk define what we know as the ring system of Saturn (Fig. 10.1).

So, the protosolar nebula becomes a disk with a series of gaps in the surface distribution of its material, where denser rings (soon to be protoplanets) circle, steadily gaining mass as particles spiral inwards, from outside, and outwards, from inside, to enhance the density of the ring. A set of well-separated planets can result from these essentially "tidal" mechanisms. In time, the protostar is "de-spun", but its orbiting planets act to conserve the system's angular momentum. We have formed a planetary system as a natural consequence of the star's need to slow its rotation.

10.3.   The timescale of planetary formation

Whether we can detect such a disk depends on many factors: how much total cloud material is available; how much matter winds up in the disk as opposed to the infant stellar core; how long it takes planets to form, thereby depleting the supply of circumstellar dust particles.

This latter issue is a vital one with suggested answers varying from a brief 100,000 years to a tedious few billion years for a central star that will ultimately resemble our sun. If planets really form rapidly then our chances of observing a substantial dust disk are a thousand times poorer than if the more sluggish process operates. Where do we look to test these estimates of time scale? We might look to mature hydrogen-burning stable stars like our sun: objects in middle age – 5000 million years for the sun – with seething atmospheric surfaces whose apparent vigour is a mere ghostly echo of the activity of the formative years. Alternatively we could investigate the population of visible young stars that will grow into solar-type stars but which are now seen at ages from around 500,000 to a few million years. These are the T Tauri stars, of course whose properties have served to focus our ideas on low-mass star formation. Finally, suppose that we could isolate a group

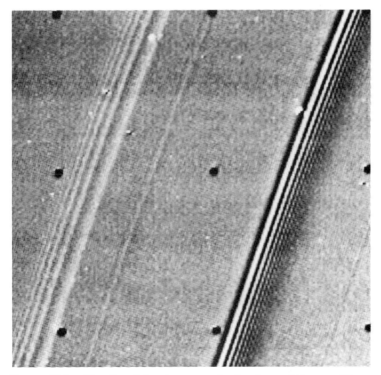

Figure **10.1**   Details of density waves in the rings of Saturn. (Voyager photograph, kindly supplied by Jet Propulsion Laboratory.)

of objects so young that they were the precursors of even the young T Tauri stars. These would be closer to the epoch of star birth and might be expected to reflect the intimate details of the fabrication of the dust shells around young stars. Precursors would certainly be no older than 100,000 years and often could be only a few tens of thousands of years old. As candidates we would, of course, propose the stars that excite HH nebulæ (Chapter 6).

10.4.   Are there much older stars with dust disks?

What do we learn from mature stars? If we viewed our solar system from far outside its boundaries we would discover that the sun too has an excess of infrared radiation, over what such a star would be naively expected to produce. But this excess is a mere 1% of the total solar radiation and it reflects the greatly diminished amount of dusty material that still litters the solar system. This matter failed to become permanently incorporated into larger chunks such as planets, moons or comets, and now it remains, circling the sun between and beyond the planets, in a very tenuous ring that would be extremely difficult to detect from far away, even in the infrared. Nonetheless, mature stars have recently become newsworthy items at infrared wavelengths. It was due to the Infrared Astronomical Satellite that made very long wavelength (sensitive to very cool dust) measurements of a number of mature, nearby, optically bright stars. Most remarkable is Vega, a star much hotter than the sun, that possesses a very cool dusty zone whose dimensions can be measured by IRAS (a similar zone surrounds Beta Pictoris). What is seen is thought to represent a ring of cool (85 K) particles, each larger than one millimetre in radius, forming a structure twice as large as the orbit of Pluto. This dust must be left over from the birth of Vega and may represent a distant region of debris akin to our solar system's comet-forming zone. Another mature star, Epsilon Eridani, has such a substantial cool debris cloud that comets must be very frequent events in its inner planetary system (if it has one). None of these observations bear directly upon the existence of planets but they do suggest a kinship with our outer solar system. Further, these stars show no evidence of hotter dust that could exist close to their surfaces, at planetary distances. Either there once was hot dust that has now accumulated into inner planets or perhaps there was never an appreciable mass of hot material to form rocky planets.

Another relevant and recent story relates to the disk observed optically around the main sequence star, Beta Pictoris, a star with twice the sun's mass. Here, by careful optical scrutiny with a new imaging sensor, it has been possible spatially to resolve a large, fuzzy, flattened cloud of circumstellar material. Attention was re-directed to this star because of IRAS's discovery of a shell of anomalous cool dust emission, just as for Vega.

Looking along the main sequence, we find an interesting phenomenon associated with stellar rotation. All stars a little more massive than the sun (around 1.5 solar masses) arrive at the main sequence with substantial angular velocities (the most massive with speeds in excess of 450 km/sec) whereas stars below this mass are slow rotators (less than about 25 km/sec). Now, among the precursors of these stars – the T Tauri stars that will one day reach the main sequence with masses greater than 1.5 solar masses – we also find relatively fast rotators (Chapter 5.6). However, those with masses below 1.5 solar masses are already slow rotators during their T Tauri phase. Perhaps we can identify the slow rotators on the main sequence with stars that have "solved their angular momentum problems" by producing planetary systems. Likewise, the fast rotators may have failed to generate planets to carry away the stellar spin. What is intriguing about the spin of the T Tauri stars is the implied timescale for the formation of planets. The essential difference between the pre-main-sequence evolutionary tracks of T Tauri stars with mass above and below 1.5 solar masses (see Fig. 5.9) is the existence and length of the convective tracks. Higher mass objects evolve essentially immediately onto radiative tracks while the limiting case, of too low a stellar mass (below about one third of a solar mass), yields a track that is always convective.

Could there be a process that operates along the convective tracks that causes stars to be spun down? The timescale for forming planets, if this were the relevant mechanism, must be less than around a million years for stars like the sun. This is much more rapid than claimed by some of the theories that have been proposed for the slow, gradual accretion of the planets. In some of these models, planets form over as long as 200 million years.

There is another clue to the rapidity of the planet-forming mechanism. This comes from the remarkable star, HL Tau.

10.5. HL Tauri and its disk

As we have already discovered in this book, T Tauri stars are so much closer to the times of their births than are ordinary stars that much attention has focussed on them and, in particular, upon the optically faint star, HL Tau. HL Tau is the only one of these stars that reveals the presence of the fingerprint of circumstellar water-ice grains in absorption towards it (Fig. 10.2); it has the deepest silicate dust absorption feature among the class; it is the most strongly polarised of these stars at visible and near-infrared wavelengths. Put all this together and what do you have? These extreme properties suggest that we view this one T Tauri

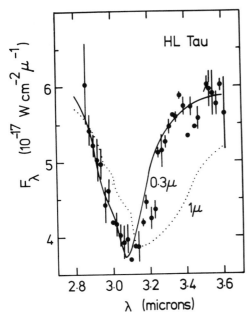

Figure 10.2   Infrared water-ice signature in the spectrum of HL Tau showing how well the observations are fitted by laboratory data for 0.3 micrometre radius ice particles.

star directly through the disk plane of its formative solar system.

Now, what we can do with these absorption features is to estimate the mass in the form of small particles necessary to produce the ice and silicate fingerprints. This amounts to a very tiny dust mass. One aspect of observations of dust to keep in mind is that, at any specific wavelength, we are principally sensitive to thermal emission from grains over a narrow range of sizes, roughly comparable to the wavelength. So if we fail to detect any small grains we should increase the wavelength of our observations and seek somewhat larger particles. Even so, the amount of material that is observed by airborne infrared techniques, that radiates at long wavelengths (like Vega's debris), is minimal, and estimates of the mass of the observable dust cloud are a tiny fraction of the mass of our solar nebula. Multiplying this dust mass by the usual ratio of gas-to-dust masses in the interstellar medium (about 100), since gas (mostly hydrogen) is always much more abundant than dust (essentially the heavier, cosmically less abundant, elements) in space, still yields a total mass in material that is beneath what we might term "the minimum solar nebula". This is the mass of stuff that our own solar system has, in the form of planets, asteroids, comets and more widely distributed debris (the interplanetary medium of gas and "zodiacal" dust: so-called because its actual spatial distribution appears projected through the constellations of the zodiac). If, and this could be a big and unjustified assumption, HL Tau has any relevance to planet formation and to solar systems like our own, then even by its early age (only 100,000 years) it has disposed of its small particles and gas. Is this vital evidence that planets have already formed around this remarkable star in only 100,000 years?

Where might these materials have gone? Perhaps the vigorous polar jets that characterise the pre-T Tauri phases have "vacuumed clean" the disk of all lightweight debris. Planets, of course, would not have been swept away by these earnest vacuum cleaners, nor (probably) accreted onto a protostellar core. It is, therefore, very tempting and perhaps even correct to conclude that HL Tau has already formed its planets. By extension, perhaps this very early planet-building characterises many other low-mass T Tauri stars which are slow rotators.

There is a tantalising clue from airborne measurements that the coolest dust around HL Tau is organised into an extended and measurable region, perhaps a hundred times that of Pluto's orbit, but too small for IRAS to see as more than a point at the distance of this star (500 light years). The high optical polarisation of HL Tau speaks for a high degree of organisation of the circumstellar dust as starlight must be scattered into our direction from a rather asymmetrical cloud. On much smaller scales (only about 4 times the orbit of Pluto), very fine-scale infrared images of HL Tau also suggest that it is elongated. Again, individual planets cannot be seen, unless they are super-Jupiters in size, but the existence of an extended, flattened dusty nebula is plain.

10.6. Other infrared observations of disks

Most remarkable are the unprecedentedly vigorous and well-collimated phenomena that characterise stages of stellar life even earlier than HL Tau; these formed the subject of the latter portions of Chapter 6. The characteristic is the presence of an infrared-emitting dust cloud like a ballerina's tutu, perpendicular to the axis of ejection of material from the central star. The phenomenon is replayed by several extremely young systems where airborne observations distinguish cool dust structures unresolved in the direction of flow from the central stars but extended perpendicular to the sense of flow. A striking example of this (Fig. 6.9a) is an infrared object associated with a visible, one-sided family of no less than five distinct blobs of gas, thrown off repeatedly (perhaps even periodically?) by the hidden star at the end. Airborne data reveal a well-resolved structure resembling a disk, of diameter almost 130 times Pluto's orbit, viewed close to edge-on, and orientated perpendicular to the chain of gaseous missiles. All these findings can be summarised into a tantalising sequence. When extremely young, stars have disks of substantial diameter that contain spinning dusty debris from the business of starbirth. In time, the disk thins and shrinks in radius as material spirals into the growing central star-to-be, or is carried away from the stellar vicinity by powerful collimated jets from the poles of the star, or clumps into protoplanets. If we are very fortunate we may view such an infant solar system from its disk plane and can observe the matter resident in the disk; but so thin are these disks that the alignment required for this is critical, hence the rarity of HL Tau among its peers.

10.7. The observability of planets

Is there any direct way to observe all these putative planets and to test the provocative idea of rapid planet-formation? Suppose you have a fixed amount of solid material and you want to make it most detectable in the infrared. Your best bet is always to put it in the form of a horde of small particles rather than one, or even a few, much more massive objects. Finely divided dust has a far greater radiating area than the surface area of a few big planets. This accounts for the ease of detection of circumstellar dust clouds. Unfortunately it also implies the severe difficulty of detecting planets by their thermal emission. Once you look outside your own solar system, not only is the intrinsic infrared emission (and reflected starlight) of extrasolar planets very small in absolute terms, but it is swamped by the emission of the associated stars.

Suppose we were on a planet circling a star far from our present solar system. Could we detect any of the sun's retinue? Jupiter is the dominant planet of our solar system and we shall use it in what follows to exemplify the difficulties of detecting extrasolar planets; adding the effects of Saturn would not help and no other known planet would be any easier to see from far away.

At visible wavelengths the situation is hopeless. Even were a planet a perfect mirror, so dilute is the intensity of sunlight at the distances of the planets that it would constitute only a minuscule fraction of the combined light of star and planet(s). In fact, the reflected light even of Jupiter cannot exceed 3 parts in 1000 million of the brightness of the sun. So much for the photometric (light-measuring) precision necessary to detect even Jupiter from outside our solar system! Another way of detecting planets would arise were we lucky enough to view the solar system from its "ecliptic plane" in which lie most of the planetary orbits. That would mean that once every twelve years we would see Jupiter pass in front of the sun's disk, at which time it would block out 1% of the sunlight for a time of the order of 2 days. We would need sharp eyes, great patience and good luck in our selection of viewing direction, but the small separation of Jupiter from the sun, and its tiny apparent size, would both be irrelevant for this experiment.

What about taking a photograph or (better) an electronic image of Jupiter, that might show it separately from the sun (assuming that our camera could record something as faint as 3/1,000,000,000ths of the sun's apparent brilliance, given the likely presence of the sun in the same camera field)? Were we stationed even at the distance of the next nearest star to earth, we would view Jupiter and the sun from more than 4 light years! The angular separation between Jupiter and the sun would be only 3 seconds of arc (the angle subtended by a 1p coin at a distance of one kilometre!). This might be a feasible experiment from a spaceborne platform, with the major limitation set only by the relative faintness of planet and sun.

Jupiter's temperature is about 100 K. This is not dictated purely by incident sunlight that warms the planet's surface. As we mentioned in Chapter 2.7, Jupiter is a brown dwarf. As such, it is making an effort to contract, little by little, to raise its internal temperature. Although it will never achieve hydrogen-burning, it is shrinking very gradually (about 0.4 mm per century!) and this gravitational contraction does liberate energy which the planet radiates at long infrared wavelengths. In fact, Jupiter's internal heat source provides about three times as much thermal output as the incident sunlight. Where in the electromagnetic spectrum does Jupiter radiate most of its energy? At about 30 micrometres (60 times the wavelength of yellow light), at which wavelength the contrast between Jupiter and the sun is at a minimum. But don't get your hopes up too high. At visible wavelengths Jupiter reflects less than 3 parts in a thousand million of the sun's light; at 30 micrometres it still emits only 8 millionths of the sun's infrared emission, even though the infrared situation is two thousand times better than the optical one! Once again there is no way of achieving sufficient photometric precision at 30 micrometres to test for such a Lilliputian excess of brightness over pure solar radiation. Our best shot, even in the far-infrared, would be to achieve sufficient spatial resolution on an ultrasensitive imaging device in space to separate

Jupiter from the solar disk. Given a spaceborne infrared telescope of aperture only 100 inches (2.5 metres) this experiment could in principle be carried out. Of course, inverting the situation, nobody's promised us that there is a Jovian-sized planet circling Proxima Centauri, our nearest known extrasolar neighbour. If we had to go to 10 light years to find a suitable extrasolar planet this would necessitate a spaceborne 200 inch telescope operating at far-infrared wavelengths or perhaps an aperture synthesis instrument consisting of several smaller infrared telescopes working as does the VLA (Chapter 6.7). However, this technique may well offer a reasonable chance of success, subject only to the giant uncertainty of knowing at which star to point our planet-seeker.

How might we enhance the chance of success? Think about a binary system in which two stars orbit around one another. Where exactly is the point about which the stars rotate? It's the centre of mass of the system (Fig. 10.3). If both stars are of equal mass, we can easily sketch their orbits (Fig. 10.4). But take, for example, the earth-moon system – virtually a binary planet since the ratio of satellite to primary (moon to earth) masses is larger than for any other planet and its moons in our solar system. The earth outweighs the moon by a factor of 81 so the centre of mass of our system is actually a point 1000 miles below the earth's surface! Similarly we can ask where the centre of mass of the Jupiter-sun pair lies. Jupiter has about 1/1000th of the solar mass so their centre of mass lies about 0.4 million miles below the solar surface. Let's

a

b

Figure **10.3**   Centre of mass (cross) in two stellar binary systems: a) equal mass components; b) unequal.

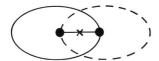

Figure **10.4**   Binary orbits in an equal mass system.

retreat again to our extrasolar vantage point to view this situation. In the absence of an infrared high-resolution satellite we are unable to see Jupiter directly but we have plotted the position of our nearest stellar neighbour, Sol, with some precision over the centuries. What do we find? Fig. 10.5 represents the path of this yellow star.

It wobbles about its mean direction. At this point we might become quite excited for we would recognise the signature of a perturbing body near the sun. Although small, the gravitational see-saw effect of some sizeable body on Sol's path is evident. From the magnitude of the wobble we would convince ourselves that it is too small to represent an undetected star (we can actually calculate the ratio of masses of the unseen perturber and Sol) but could signify a giant planet! Of course, if the purpose of our experiment were to detect a potential life-bearing planet we would have failed. But, for now, the battle is simply to provide convincing proof of the existence of *any* extrasolar planet, no matter how big. And there the matter rests for this decade. It has been proposed to construct an infrared high-resolution instrument to seek planets by their far-infrared radiation, but a very sensitive optical wobble-seeker is a potentially better candidate for a future spaceborne device.

If one were impatient to begin the search now, it would be necessary to construct a very specialised telescope, capable of

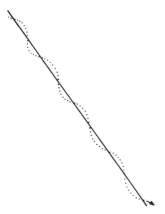

Figure **10.5**   Motion of a star perturbed by an unseen companion. Solid line represents the average direction of motion; dotted, the perturbations on this orbit.

continuously monitoring the light output of about 10,000 candidate stars with a precision for each measurement of 1 part in 1000. This could promise a yield of one giant planet per year of observing time, by detecting the tiny dimming caused by transit of the large bodies (Uranus and up in size) across their suns' disks.

It was suggested, in late 1984, that the very cool low-mass red dwarf star, van Biesbroeck 8 (vB 8), has a planetary companion. By the use of high-powered infrared techniques it has been possible spatially to resolve a very cool companion object to this already cool star. The mass of this companion is estimated to be perhaps 10 times that of Jupiter. As such, we are dealing strictly with a brown dwarf rather than with a "planet", in the terrestrial sense of the word. Of course, there is a rather fuzzy boundary between large planets and failed stars! It has long been felt by astronomers that the best chance for recognising a brown dwarf might well be to look among all the nearest stars (which sample includes vB 8) for an infrared excess or perhaps a spatially distinct, very cool component. By restricting the sample to the nearest stars in space we enhance our chances of being able to resolve potential companions from their primary stars. To this extent, vB 8's companion represents a vindication of this idea and points out the essential correctness of the idea that brown dwarfs can exist. In terms of planets, habitable by creatures in any way remotely like ourselves, this object is an unlikely place; indeed, it must be even less likely than Jupiter. Of course, there are always moons around these giant planets, with more hospitable (less-crushing) gravities...

10.8. Rings and things

It is highly legitimate to ask the question: "What have we learned from space vehicular exploration of the solar system that is relevant to our knowledge of the early solar nebula?" In fact, the replies to such a question already occupy many volumes of observational data, gleaned from NASA's, and the U.S.S.R.'s, various spacecraft over the years, and their interpretations have been fought over at numerous conferences. I will touch upon just one aspect of solar system research that bears upon the topic of this chapter. This refers to the bounty of wonderfully detailed images of Jupiter and Saturn and their ring systems, yielded first by the Pioneers, then by the Voyagers.

Images of Saturn's rings (Fig. 10.1) plainly illustrate the difficulty of establishing a dividing line between "ring particles" and "moons". Traditionally, the larger bodies that can be identified as separate entities have been called moons whereas those observable solely en masse have been termed ring particles. Clearly a continuum exists between large particles and tiny moons in the rings. There are very complex interactions between moons and the myriads of tiny objects that constitute rings, which bear upon the mechanisms of formation of the latter. Jupiter, too, possesses a ring – to be sure, it doesn't have an extensive and complicated system as do Saturn and Uranus (5 major rings exist around Uranus, but we have found a more elaborate system during the 1986 "encounter" by a Voyager spacecraft with Uranus), but it does have a definte ring. It is thought that the particles in Jupiter's ring could have been eroded from larger Jovian satellites and that these ring particles are only short-lived, before they spiral away, some into Jupiter, others far from it. We have "shepherds" (Fig. 10.6) around Saturn and Uranus – moons that literally herd together particles by their "tidal" (gravitational) forces, pushing and pulling them into very narrow ring structures. In short, it is difficult to separate rings from moons.

How did planetary rings form? Their evolution is presumed to have mimicked that of the early solar nebula. Giant, flattened debris clouds surrounded each major young planet due to conservation of angular momentum during the collapse of the protoplanetary clouds. Satellites formed within planetary disks exactly as planets did in the disk of the solar nebula. The highly complex structure that the Voyagers have revealed in Saturn's rings is largely due to the major Saturnian satellites and especially to the small "ringmoons" that are found just beyond the edges of, and embedded within, the rings. These structures evolve on rapid timescales due to the tidal effects of rings and moons on each other, which effects are mediated through spiral density waves, in concept like those that we described as causing spiral arms in our own Galaxy (Chapter 3). It is perfectly possible that collisions, or close encounters, between early moons disrupted these bodies, providing material for the planetary ring systems.

How could planetary disks bear any relation to the solar disk? Clearly, planets and their rings are physically much tinier than the early solar nebula, both in extent and total mass. Yet the protosun also derived its basic energy source from the radiation of energy gained by gravitational collapse, as do Jupiter and Saturn even now (remember that these giant planets are also "failed stars": Chapter 2.7). Although the early sun was undoubtedly hotter than the rather chilly 90-100 K of modern Jupiter and Saturn, nevertheless the distance between sun and planets greatly exceeds the orbits of

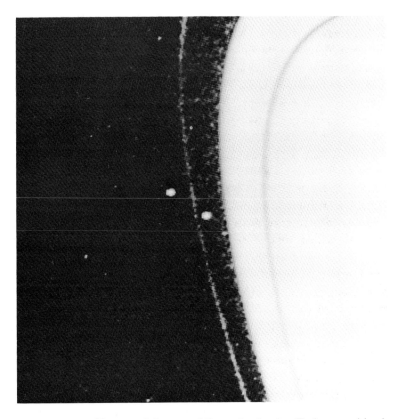

Figure **10.6** Voyager 2 image of Saturn's slender F-ring outside the bright A-ring. Bracketting the F-ring are its two "shepherd" moons. (Image kindly provided by Jet Propulsion Laboratory.)

satellites around planets. Consequently, the planets could have influenced the formation of their moons in qualitatively the same way as the sun may have affected the condensation of its retinue. In some ways then, the study of planetary rings tells us of conditions and processes relevant to planet formation. Of course, the mere existence of so many moonlets around Saturn suggests that tidal forces inhibit accretion, otherwise, by now, there would have been a single massive moon!

The present-day moons of a planet can be of two varieties: those that were genuinely made from the debris of formation of that planet, and those that represent the capture of an interloping body that just happened to be in the vicinity but was actually made elsewhere in the solar system. The first kind of moon bears upon the issue of the chemical composition of bodies made in different (but known) locations in the early solar system. Usually these satellites orbit their planets in the same sense as the planets rotate on their axes. However, captured bodies often tend to be in "retrograde" orbits; that is, they orbit their primaries in a direction opposite to the planetary rotations. These retrograde bodies are thought to have been captured prior to the final collapse of their primaries' protoplanetary clouds. Some planetologists do not believe that the inner (terrestrial) members of our solar system ever possessed substantial nebulæ, consequently they must explain our Moon and the two tiny satellites of Mars by other means than formation in a protoplanetary disk: perhaps these bodies are all captured wanderers. Of course, provided that the moons have sensibly the same chemical composition as their primaries, it is conceivable that the satellites are merely pieces broken from their planets (the "fission" hypothesis). Equally, there may have been a whole flock of moons among the terrestrial planets but their orbits could have decayed already (i.e. come in closer to their primaries), leading to their destruction. Such an idea has been advanced for at least tiny Phobos, which orbits Mars, but which will be swallowed up by that planet in only a few hundred million years. Perhaps processes such as these terminate the hierarchy of star-planet-moon and don't permit moons of moons to survive, even if they could form.

10.9. The scale of disks

Our (known) solar system at 40 AU diameter is a fait accompli. Large, cool disks with diameters of several thousand AU are turning up in far-infrared observations of the youngest stars. Why should there be such a difference in scale between these two observable classes of disk?

Firstly, theoretical disks are always in a state of flux as they evolve, with angular momentum being transported both inwards and outwards. A single distant planet is an excellent repository for substantial angular momentum. Therefore, until we have really established the outer bounds of our own solar system, no matter how tenuous, we do not know that there is any discrepancy between our planetary system and the giant, cool disks around HH-exciting stars. Second, we have seen that molecular clouds have both cores and envelopes. The collapse of the inner part of a rotating core alone will tend to produce a rather small disk, only a few AU across. However, the envelopes of clouds blend smoothly into their cores. It is natural to expect that some of this material, with relatively high angular momentum, will also try to fall into a protostar. The envelope must create a giant disk, thousands of AU in diameter.

The theoretical frontiers in this field focus on the proportions of mass that are resident in the great disks, as opposed to their small inner parts; on the question of how binary stars form (as two separate accreting centres in a common infalling envelope or as two independent entities that come together much later in their evolution); and on the interplay between potential binary protostars, their accretion flow(s), and the existence of disks. Judging from L1551 IRS5 (Fig. 6.11), binaries are clearly of interest, all the more so because, on the main sequence, the majority of stars are members of multiple systems (binaries and even more numerous families).

10.10. Conclusions

HL Tau presents us with a clue that perhaps the beginnings of planetary build-up have commenced only 100,000 years after the start of cloud collapse. Once particles are agglomerated into units the size of golf-balls or Volkswagens, it is practically impossible to observe them. Older T Tauri stars (a million or more years old) show mere vestiges of these early disks. IRAS' contribution may be the realisation that cool ghosts of these disks persist even for hundreds of millions of years into stellar evolution, their particles waiting to be locked up into comets.

We have witnessed the dismemberment and disappearance of sizeable dusty disks around young stars that will one day be very like our present sun. Unless stellar jets can effectively scour clean almost all the disk material, planets represent a very plausible repository for the "vanished" matter. We cannot yet point a finger at entities that are likely to be individual planets around distant stars but we do believe we now know those sites in which planets are being assembled and we recognise that planet-building is a process that wastes no time in early stellar life before its inception.

EPILOGUE

You can never plan the future by the past.
Edmund Burke
Letter to a Member of the
National Assembly

The only thing of which an astronomer can be certain is that next year's Universe is likely to seem quite different from this year's model. We are poised on a most exciting brink, scientifically that is. Our dependence on satellites and space-shuttle-borne instruments for access to exotic parts of the spectrum is going to increase. Infrared astronomy has just achieved its first major, unbiased survey and a range of follow-on space infrared observatories is already planned. Ultraviolet is a fully-fledged satellite discipline with a background of about a decade of remote observatories, and the Space Telescope will revolutionise this field (also that of optical astronomy) by providing deep sensitivity and fine detail through its appreciable aperture above the atmosphere. X-ray telescopes are between the highs of a few years ago, when the "Einstein" High Energy Observatory was still active, and the highs to come when new European and new American facilities will start to operate, taking over from the current European "EXOSAT" instrument. Gamma rays – the most energetic photons – are presently being observed but a bigger orbiting telescope is planned. On the ground we dare not become complacent, however. For every spaceborne observatory we still need a complement of ancillary ground-based optical, infrared and microwave telescopes. Here, too, astronomy stands at the threshold of a new era. A multi-segment 10 metre optical/infrared telescope is under construction on Hawaii. There is serious talk of an airborne 3.0 metre telescope, to be housed in a jumbo jet and to fly at substantial ceilings to grant sensitive entry to otherwise inaccessible (far-infrared) wavelength regimes. In Japan a 45 metre dish, with an accurate surface, is already making dents in high-spatial-resolution millimetre wave-length progammes. A joint Franco-German 30 metre antenna is coming on the air as I write this chapter, and, at several locations, multiple dish millimetric aperture synthesis instruments are either in operation or are planned (short wavelength arrays styled on the VLA: Chapter 6.7). At longer wavelengths, giant intercontinental networks of antennas act as a single monster array to provide superfine spatial resolution for observations of the most luminous objects in the radio Universe. In short, spaceborne astronomy should stimulate, not stifle, the growth of ground-based astronomy.

Clearly, there are almost limitless problems to be tackled even in our restricted sector of astronomy. Much future work must focus on the gaps in our knowledge of what lies between cloud cores and accreting protostars. High spatial resolution offers us an opportunity to study the evolution of the giant circumstellar disks both in molecular lines and through their dust emission. How do potential solar systems form? Do we know all the repositories of angular momentum even in our own solar system or is there another, hitherto undiscovered, giant planet (or planets) far beyond Pluto's lonely orbit? Do pre-T Tauri stars power their jets by "relativistic" means, in which electrons or protons move at speeds closely approaching the velocity of light? This is the mechanism currently favoured for the blazing beacons of quasars to drive their own collimated flows but can Nature scale down this prodigious machine to operate in a low-luminosity stellar context? Did the sun form singly or is it a member of a binary with a far-flung faint companion light years away? Are many T Tauri stars binaries? Do planets form as rapidly as we now believe and are potential planetary systems to be inferred around all slowly-rotating solarlike stars? What of high-mass stars? Could they harbour planets too? What slows down the rotation of molecular cloud cores? How important is the magnetic field both in the ambient medium around a forming star and within the confines of that star?

When we have asked, and answered, enough questions like these we may be ready for the truly cosmic issues such as our own origins, the possibility of our uniqueness or non-uniqueness, the fate of the Universe itself.

There are no real horizons in astronomy – only places where our knowledge, or perhaps our imagination, has not yet granted us entry. There are voids in our observations and in our theoretical studies but these seem always to be bridgable temporarily by judicious application of intuitive reasoning or by the recognition of morphological similarities in other astronomical areas. Later, faster computers, or more ingenious software, will render today's intractable theoretical problems tomorrow's success stories. Likewise, new technology breeds new sensors with astronomical applications that can attack observational challenges that once seemed insuperable.

Few emotional highs can compare with the thrill of playing a hunch successfully. Even a failure to detect, or to demonstrate, some logically predicted phenomenon has value, for it constrains the possible behaviour of some piece of the Universe. In this patient, and sometimes breathless, probing of the cosmos lies the essence of good research. If you have gained some sense of how we pursue our interests in star formation, then this book will have achieved at least one of its aims.

INDEX

by Chapter Section